ADVANCED TESTING TECHN
A PRACTICAL GUIDE FOR ELEC
FIRST EDITION
By Anthony Hinsley MA, MIET
COPYRIGHTS, ACKNOWLEDGEMENTS

Text copyright © 2007 Anthony Hinsley.
Tables © 2007 Anthony Hinsley.
Pictures drawings and layout © 2007 Anthony Hinsley.
First Edition published and printed in France 2007.

Every effort has been made to ensure the information and guidance in this book is correct and accurate at the time of going to press. However, it is understood that all parties must use their skill, knowledge, experience and understanding when making use of it.

Neither the author, publisher or any named person hold themselves responsible for any consequences that may arise from errors or omissions and negate any liabilities.

Permission to reproduce extracts from the BS 7671(2001): 2004 is granted by BSI. British Standards can be obtained from BSI Customer Services, 389, Chiswick High Road, London W4 4AL. Tel: +44(0)20 89969001. email: cservices@bsi-global.com

The drawings of instruments in this book are graphical representations of concept meters and are used so as not to reflect any particular manufacturer, make or model, allowances must be made for their realism.

I would like to thank Sally, P & R Hurt Education & Training and Nigel Bond (Kier Group) for their input.
Published by
Castleknight Publications

ISBN13: 978-2-9524138-0-0
ISBN10: 2-9524138-0-0

CONTENTS

ADVANCED TESTING TECHNIQUES
A PRACTICAL GUIDE FOR ELECTRICIANS

Introduction

To introduce myself, I am a qualified Electrician with more than 25 years experience in the industry, now writing and supplying resources to the electrical training industry. This book is for electricians who want to have a better understanding of the concepts of testing techniques, and I address this book to all training and practising Electricians.

Throughout the text I will address all my readers as "You" just as I would if we were in verbal discussion together as I believe that this will make for easier reading and understanding.

This book is aimed at Electrician's who would like to extend their testing knowledge and ability especially in a maintenance environment.

I will be looking at the more complex testing methods to create a better understanding, for instance, do you know how to fault find on control circuits in their de-energised state? Do you know how to determine what components are faulty in discharge lighting circuits?

If your answer to either of these questions is no, then I hope that after reading and putting into practice the tests in this book, you will be confident that you have the necessary competence to carry out the tests effectively and safely.

Of course with all new skills, one has to practice over a period of time to become proficient.

As you progress through this book you will gradually realise that all types of testing can be carried out more safely by testing in safe zones and with the circuits and equipment in their de-energised state.

Parts of this book are similar to my first book Testing Electrical Installations A Practical Guide For Electricians, this is because **SAFETY** is paramount when dealing with electricity so some Health and Safety issues have been repeated.

THINKING OUTSIDE THE BOX

All testing procedures have a degree of danger associated with them, and it's up to you the person carrying out the tests to ensure that yours and everyone's safety is maintained during all phases of any test procedure.

You can also minimise the risks associated with electrical tests if you are indeed competent, this means that you possess the necessary knowledge, skills and experience to carry out your job function.

You should also have a sound knowledge relevant to the installation or part of being inspected, tested and worked upon.

You will need to carry out a risk assessment which will involve a detailed inspection of the installation and or circuits to familiarise yourself with all of the various functions of that installation or circuit.

Shutting down the wrong circuit could inadvertently cause danger, definatly cause dissruption to the working day and may cause loss of important data processing, take your time to think carefully and try to anticipate the outcome of your actions.

You should be aware that once engaged in testing you will need to concentrate on the task in hand from start to finish, distractions cause complacency and create danger, it is recommended that you have a good working knowledge of safe working practices as detailed in HSG85. (see page 79 note: 3)

The risk of electric shock is not just the product of touching a live part within an installation, it can be brought about during the testing procedures by inadvertently inducing voltages on exposed conductive parts of the installation.

The tester is also at risk from electric shock due to un-discharged capacitive loads in a piece of equipment such as capacitor start motors, fluorescent fittings, and capacitance induced in to cable systems during insulation testing.

Fault finding puts the tester under pressure to solve the fault quickly, this in turn creates a dangerous situation, and you must stay calm, be methodical in your approach, think laterally and make sure that the piece of equipment is fully and safely isolated.

You must make a thorough assessment of your surroundings to maintain safe working practices.

Check and double check before proceeding...............

GATHERING INFORMATION

Gathering information about the job or task to be undertaken is fundamental in helping towards a professional attitude and successful completion of the job or task. Before you leave for the job, you need detailed information relating to the type and use of the premises and any special precautions that you may need to take, such as protective clothing.

Will you need to ask for someone to gain access to the premises? Get their name, telephone extension number and any other relevant information, remember to give them your details so they know who's coming, and in what vehicle, if necessary give a registration number.

Will you be able to gain access with your mobile phone and other sensitive equipment? If you can gain access with mobile phones and other sensitive equipment is there any part of the premises that you can not use the equipment?

Don't forget that you may be asked to prove who you are. Take the relevant information, preferably something with your picture on it.

Once you arrive on site, contact the person responsible for the installation or at least the person who knows where relevant services, circuits, etc can be turned off e.g. the location of the main switch room. You will also need;

Diagrams, charts, tables and any relevant documentation for the installation regarding:
- Isolation methods and their locations.
- Type of earthing arrangements.
- Safety supplies such as back-up generators and UPS systems.
- Any circuit or equipment vulnerable to isolation or a particular test.

However when checking records, care should be taken not to rely on one source of information, as records are sometimes out of date or incorrect, so carryout an installation survey.

One of the key points regarding testing and fault diagnoses is to extract the right information from the person operating the equipment. Talk to them asking probing questions such as where, when and what happened. Some faults can be associated with other equipment and don't necessarily relate to the account given by the operator. Always keep an open mind and think laterally, consider everything and discount nothing no matter how trivial.

TEST EQUIPMENT

Let's discuss briefly the instruments that I will be dealing with.

Insulation resistance and low resistance ohmmeter

The instrument allows the measurement of the resistance in a circuit in Ω's and can generate voltages to detect insulation leakage between live conductors and live conductors to earth. Continuity testing at open circuit voltages between 4V and 24V ac or dc and able to deliver a closed circuit current of 200mA for this purpose.
Insulation testing at voltages of 250V, 500V and 1000V dc, at 1mA.

Earth loop impedance tester:

This instrument is designed to measure the resistance of the circuit utilising the voltage and can apply test currents up to 25A. When you carry out the test the instrument internally carries out calculations to display the resistance in ohms. This instrument also allows the calculation or measurement of **PSCC (prospective short circuit current)** Phase to Neutral and **PEFC (prospective earth fault current)** Phase to Earth. You should also have a special lead set to enable you to test circuits other than socket outlets.

Current clamp meters including low current detection of 0.01mA:

This instrument is designed and used to measure the current flowing through a conductor, and can be used for detecting leakage currents in systems or circuits, this is carried out by measuring the magnetic field generated around the conductor when current is passing through it.

Earth testers:

This instrument is designed and used to measure the earth resistance, including continuity or ground electrode measurement using 'dead earth' method. A current of steady value is passed between an earth electrode under test and a current test spike with an additional potential spike midway. The volt drop between the outer electrodes is recorded and divided by the current passing thus allowing measurement of the earth resistance.

HEALTH AND SAFETY

Before you begin testing the equipment, installations or individual circuits, there are safety issues that need to be addressed.

BS7671 states 'Precautions shall be taken to avoid danger to persons and to avoid damage to property and installed equipment during inspection and testing'. (711-01-01)

This regulation is very broad and covers just about every conceivable occurrence that could occur. Lets look at the regulation in two parts, whilst thinking about the following points before carrying out testing:

      1] 'Precautions shall be taken to avoid danger to persons'.

• 'Have you checked all supplies in the vicinity of the job?'
• 'Have you isolated the circuit correctly?'
• 'Are you using long test leads that someone could trip over?'
• 'Have you informed people of the possibility of shock whilst you are carrying out insulation testing?'
• 'Have you checked to see if any essential services are fed from the board or circuit you intend to test, such as emergency lighting, fire alarms, life-support equipment in hospitals, gas monitoring systems, standby generators, UPS systems etc?'
• 'Have you discharged any capacitors in the circuit?'
• 'Are your instruments of the appropriate type to be used in testing in any particular environment, i.e. are you using intrinsically safe test equipment in potentially explosive areas?'

      2]'Avoid damage to property and installed equipment during inspection and testing'.

• 'Is there an RCD (Residual Current Device) protecting the circuit?'
• 'Are there computers on line?'
• 'Is there electronic equipment in the circuit?' (see note: 10 page 79)
• When removing covers to equipment 'are there any special gaskets that may need to be replaced?'
• 'Are all the loads disconnected?'
• 'Are there any mixing, heating, ultra violet or other forms of processing that need consideration for the length of time that it may be disconnected?'
• 'Is air conditioning or extraction essential to the working environment?'

This page gives an example of the sort of questions you should be asking yourself...
Remember **SAFETY FIRST!**

SAFE ISOLATION PROCEDURE

All Electrical supplies are potentially very dangerous and can kill human beings and therefore should be treated with the greatest respect.

In the following isolation procedure, I will be dealing with isolating a 3-phase circuit, so that routine inspection and test for the completion of a maintenance schedule, can be carried out on a 3-phase induction motor.

In Fig 1 on page 7, there is a typical layout of a 3 phase installation, if we now look at the items of equipment in terms of what needs to be isolated, this may help to clarify what and how the procedure should be carried out.

Firstly the main switch controls the bus-bar and all the circuits it supplies, therefore we can use the main switch to isolate the whole installation and the three isolators above the bus-bar isolate various distribution boards located around the building. (Distribution boards should be located near to the load centres to minimise long cable runs for the final circuits)

In this instance we want to isolate motor **F,** so we should not isolate at point **A** (main switch), or at point **B** (distribution board isolator), or at point **C** (isolator for all equipment being supplied by the distribution board), as all of these points will isolate more than one circuit.

However, we can isolate at point **D** or **E.** It is my recommendation that both of these points are used to make absolutely sure that under no circumstances the supply be re-instated whilst work is being carried out. (see note: 5 page 79)

Many distribution boards in current service do not have their circuits clearly identified and labelled so it can be fairly difficult to isolate an individual circuit or piece of equipment.

As you are the person responsible for confirming that equipment to be worked on is isolated correctly (in accordance with BS 6423) before work can commence, careful and thorough examination of the installation is needed to locate all of the necessary equipment and its' associated cables.

If it is impossible to locate the exact equipment concerned it may be necessary to switch off various isolators until the equipment has been located. Advanced warning must be given to the correct personnel of your intentions so that essential equipment can be shut down correctly avoiding danger and inconvenience.

Use a paperbased isolation procedure, see example certificates and notice on pages 80-82.

Understanding isolation procedures in 3 phase circuits.

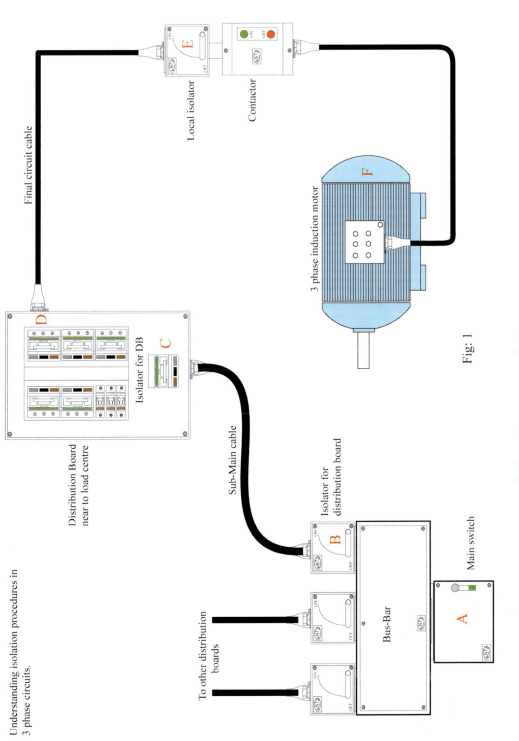

Final circuit cable

Local isolator

E

Contactor

ON

OFF

ON OFF

F

3 phase induction motor

D

Isolator for DB

C

Distribution Board near to load centre

Sub-Main cable

Isolator for distribution board

B

ON

OFF

To other distribution boards

ON

OFF

ON

OFF

Bus-Bar

Main switch

A

Fig: 1

SAFE ISOLATION OF A CIRCUIT

We have already discussed the dangers associated with supplies above extra low voltage so this procedure should be adhered to.

Refer to drawing Fig: 2 on page 9.

In brief, the sequence for isolating the motor is as follows:
- Prove voltage indicator functions by testing on a known supply or proving unit (Fig:2B)
- Locate distribution board feeding the equipment to be isolated
- Locate final circuit cables feeding the equipment
- Remove all loads (switch off and lock off or remove fuses)
- Test for the presence of voltage at the contactor using an approved voltage indicator (Fig:2A)
- Check locking off (in the case where fuses have been removed it is now good practice to disconnect the conductors from the fuse carrier)
- Test circuit/circuits are dead. Test between P-N, P-E, N-E for 1 phase and L1-L2, L2-L3, L1-L3, & L1-N, L2-N, L3-N, L1-E, L2-E, L3-E for 3 phase (See Fig: 2A)
- Test voltage indicator on known supply or proving unit (See Fig: 2B)
- Erect warning notices stating 'DANGER ELECTRICIAN AT WORK DO NOT SWITCH ON'

Remember that BS 6423 states that all **live conductors must be isolated** before work can be carried out. As the neutral conductor is classified as a live conductor this should also be disconnected to comply with the British Standard. This may mean removing the conductor from the neutral bar in distribution boards, as the neutral is not always isolated by the isolator, so the connecting sequence for neutral conductors needs to be verified and maintained. (see page 79 note 5)

You should consider the use of a paper based system incorporating, isolation certificate and "permit to work or clearance certificate" for all isolation procedures. On the warning notice 'DANGER ELECTRICIAN AT WORK DO NOT SWITCH ON', note where you are working, your name, the time and date you started work. This will help anyone to establish why power has been turned off and for how long.

All too often warning notices are left on switchgear inadvertently.

Work can now commence on all equipment associated with the isolated circuit, including the cable, contactor, local isolator and the motor.

Use a paper based isolation procedure, see example certificates and notice on pages 80-82.

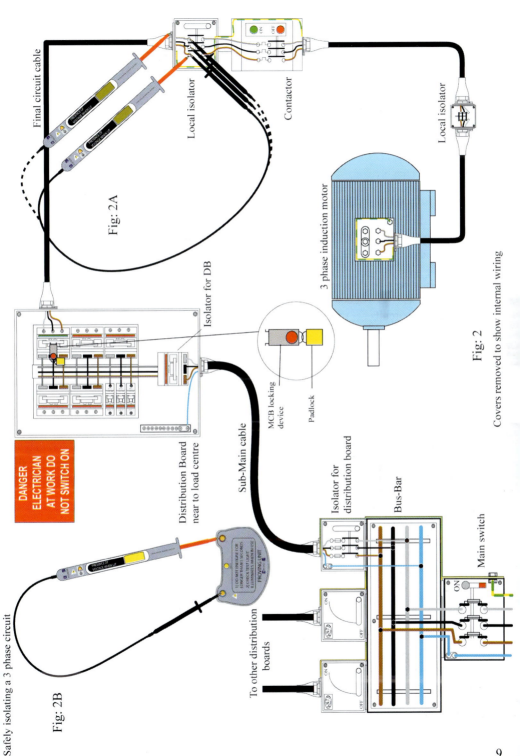

Final circuit cable

Local isolator

Contactor

Fig: 2A

3 phase induction motor

Local isolator

Fig: 2

Covers removed to show internal wiring

Isolator for DB

MCB locking device

Padlock

DANGER ELECTRICIAN AT WORK DO NOT SWITCH ON

Distribution Board near to load centre

Sub-Main cable

Isolator for distribution board

Bus-Bar

Main switch

To other distribution boards

Safely isolating a 3 phase circuit

Fig: 2B

9

SAFE ISOLATION OF A MOTOR

We have already discussed the dangers associated with supplies above extra low voltage so this procedure should be adhered to.

Refer to drawing Fig: 3 on page 11.

In brief, the sequence for isolating the motor is as follows:
- Prove voltage indicator functions by testing on a known supply or proving unit (Fig:3B)
- Locate distribution board feeding equipment to be isolated
- Locate final circuit cables feeding equipment
- Remove all loads (switch off and lock off at the local isolator and/or remove fuses)
- Test for the presence of voltage at the motor using an approved voltage indicator (Fig:3A)
- Check locking off (in the case where fuses have been removed it is now good practice to disconnect the conductors from the fuse carrier)
- Test motor is dead, test between L1-L2, L2-L3, L1-L3, & L1-N, L2-N, L3-N, L1-E, L2-E, L3-E for 3 phase (See Fig: 3A)
- Test voltage indicator on known supply or proving unit (See Fig: 3B)
- Erect warning notices stating 'DANGER ELECTRICIAN AT WORK DO NOT SWITCH ON'

Remember that BS 6423 states that all **live conductors must be isolated** before work can be carried out. As the neutral conductor is classified as a live conductor this should also be disconnected to comply with the British Standard. This may mean removing the conductor from the neutral bar in distribution boards, as the neutral is not always isolated by the isolator, so the connecting sequence for neutral conductors needs to be verified and maintained. (see page 79 note 5)

You should consider the use of a paper based system incorporating, isolation certificate and "permit to work or clearance certificate" for all isolation procedures. On the warning notice **'DANGER ELECTRICIAN AT WORK DO NOT SWITCH ON',** note where you are working, your name, the time and date you started work. This will help anyone to establish why power has been turned off and for how long.

All too often warning notices are left on switchgear inadvertently.

Work can now commence on the motor.

Use a paper based isolation procedure, see example certificates and notice on pages 80-82.

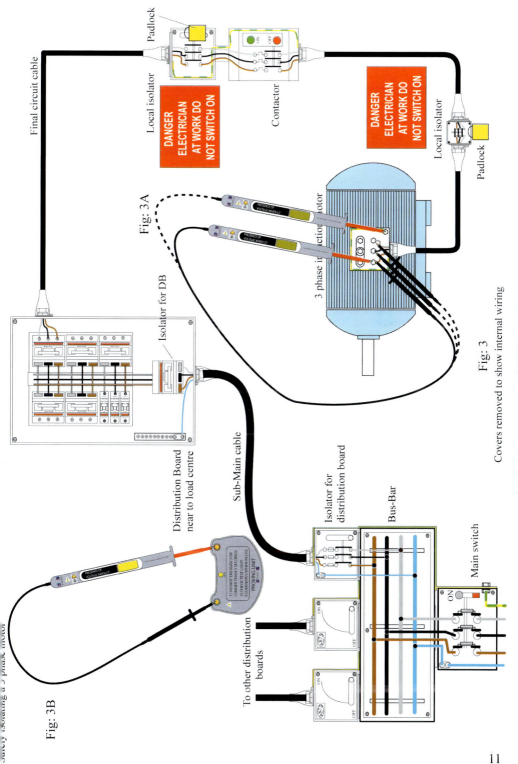

Fig: 3

Fig: 3A

Fig: 3B

Covers removed to show internal wiring

VISUAL INSPECTION

Before any metered testing takes place it is extremely important to carry out a visual inspection of the equipment and or the circuits under test.

It's your duty as an electrician to verify that the equipment complies with British Standards and or European Harmonisation Standards appropriate to the intended use of the installed equipment, and that the installed equipment isn't visibly damaged or defective so as to impair safety, and that the inspection shall include the checking of the following items where relevant.

- Connection of conductors
- Identification of conductors
- Routing of cables
- Selection of conductors
- Correct connection of equipment
- Methods of protection against electric shock
- Prevention of detrimental influences
- Presence of appropriate devices for isolation and switching
- Presence of under-voltage protective devices
- Presence of earthing conductors including main, equipotential and suplimentry bonding
- Choice of setting of protective devices
- Labelling of protective devices, switches and terminals
- Selection of equipment appropriate to external influences
- Access to switchgear and equipment
- Presence of warning signs and danger notices
- Presence of diagrams, instructions, and similar information
- Erection methods

Remember if any of the visual inspection checks require you to remove covers then you will need to carry out safe isolation, otherwise you will contravene the Electricity at Work Regs 1989.

The key point with all electrical work is that you maintain your own and everyones safety when carrying out such work.

Use a paper based isolation procedure, see example certificates and notice on pages 80-82.

Visual inspection on a 3 phase circuit

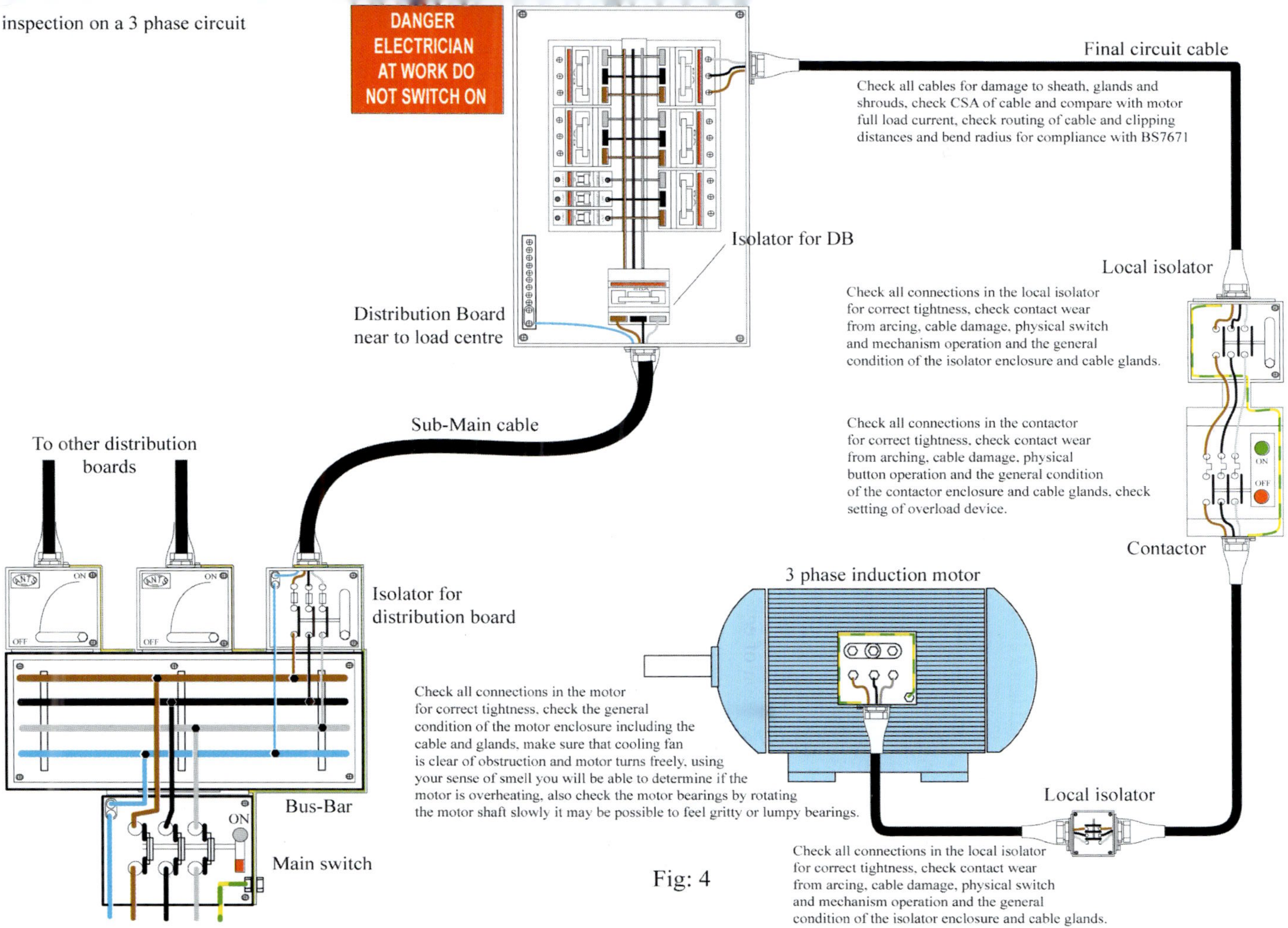

Final circuit cable

Check all cables for damage to sheath, glands and shrouds, check CSA of cable and compare with motor full load current, check routing of cable and clipping distances and bend radius for compliance with BS7671

Local isolator

Check all connections in the local isolator for correct tightness, check contact wear from arcing, cable damage, physical switch and mechanism operation and the general condition of the isolator enclosure and cable glands.

Contactor

Check all connections in the contactor for correct tightness, check contact wear from arcing, cable damage, physical button operation and the general condition of the contactor enclosure and cable glands, check setting of overload device.

Local isolator

Check all connections in the local isolator for correct tightness, check contact wear from arcing, cable damage, physical switch and mechanism operation and the general condition of the isolator enclosure and cable glands.

3 phase induction motor

Check all connections in the motor for correct tightness, check the general condition of the motor enclosure including the cable and glands, make sure that cooling fan is clear of obstruction and motor turns freely, using your sense of smell you will be able to determine if the motor is overheating, also check the motor bearings by rotating the motor shaft slowly it may be possible to feel gritty or lumpy bearings.

Fig: 4

DANGER
ELECTRICIAN
AT WORK DO
NOT SWITCH ON

Isolator for DB

Distribution Board
near to load centre

Sub-Main cable

Isolator for
distribution board

Bus-Bar

Main switch

To other distribution
boards

13

TESTING YOUR METER

It's no good just unpacking your meters and using them, it's important that you regularly check your instruments to make sure they are in good and safe working order.

You must make sure that your instruments have a current calibration certificate, and that you have ongoing confidence in the accuracy of your instruments, otherwise any tests made and recorded could be void, or inaccurate. If you are looking for definitive values and your instruments are not in good working order it could lead you to the wrong conclusions when undergoing fault diagnosis.

Depicted in Fig: 5 page 15 are a few tests that should be carried out to verify that the instrument is safe to use.

The sequence is as follows:
- Check meter for damage
- Check batteries are in good condition
- Check that the leads and probes are not damaged
- Check that the probes are of the appropriate type (GS38)
- Carry out open circuit test on all ranges to prove function of meter
- Carry out closed circuit test on all ranges
- Zero instrument on ohms scale

As the above sequence only tests the scales of the instrument and test leads to their maximum and minimum values, you should test Ω and $M\Omega$ scales with different value resistors such as 1Ω, 5Ω, 10Ω, 100Ω, and $0.5M\Omega$, $1M\Omega$, $2M\Omega$.

You should also test your RCD tester on a reference RCD protected socket. Your earth loop and voltage tester for accuracy on a known reference test socket and or source, and above all you must record these results in a monthly log as proof of on-going confidence in the accuracy of your instruments. However, allowances must be made for the tolerance of resistors and terminal voltage on batteries.

As you may know, these instruments are very costly to replace so it is important that they are maintained in good working order and that they are looked after.

If a meter is dropped or inadvertently connected to a live supply it should be thoroughly checked and tested, if however you are in doubt about the accuracy of any of your instruments you should send them for re-calibration, as this confirms accuracy/tolerance. Calibration dates and serial numbers must be recorded on test documentation.

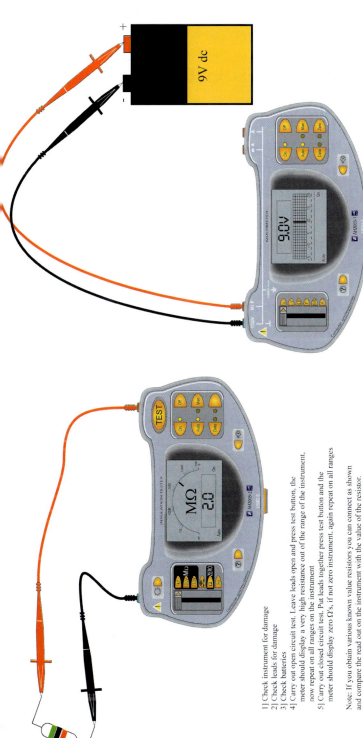

1] Check instrument for damage
2] Check leads for damage
3] Check batteries
4] Carry out open circuit test. Leave leads open and press test button, the
    meter should display a very high resistance out of the range of the instrument,
    now repeat on all ranges on the instrument
5] Carry out closed circuit test. Put leads together press test button and the
    meter should display zero Ω's, if not zero instrument, again repeat on all ranges

Note: If you obtain various known value resistors you can connect as shown
and compare the read out on the instrument with the value of the resistor.

Note: If y use a battery to indicate voltage please bear in mind
that the terminal voltage may be greater or lower than that stated on
the case of the battery. You should use a stable accurate voltage
source to verify voltage readings. The example I have given is only
to demonstrate what to do.

Fig: 5

15

TESTING 3 PHASE INDUCTION MOTORS

Firstly make yourself familiar with the motor terminals. Before you disconnect any cables or terminals make a sketch of how the wiring is connected, including writing down the terminal numbers and how the links are connected, apply the correct labelling if necessary.

Most standard 3 phase induction motors carry the numbering system for the terminals as indicated in Fig: 6 on page 17.

The configuration used in Fig: 6A shows the windings crossing over each other, when the motor links are in the vertical it can be run in delta, if the links are in the horizontal position the motor can be run in star.

Note: A motor running in star at 400V 3 phase can not be connected to run at 400V in delta, it would have to be run at 230V 3 phase in delta, otherwise the motor windings will be damaged.

Sequence for testing:

First test: Set your insulation tester or multi-meter to the highest ohms scale, zero and connect to the motor across U1 and U2, or V1 and V2, or W1 and W2, as indicated in Fig: 6A, press the test button and record the value, if no value is present change the scale until a reasonable value can be recorded.

Now carry out the same test on the other two windings. The resistance value of all three windings should be substantially the same as one another. (i.e. 1.33Ω)

Second test: Now place the links in the horizontal position as shown in Fig: 6B and using your instrument test between U2 and W2 or W2 and V2 or U2 and V2, the reading should be double that of the first test.

Now carry out the same test on the other terminals. The resistance value of all three terminals should be substantially the same as one another. (i.e. 2.66Ω)

Third test: Now place the links in the vertical position as shown in Fig: 6, and using your instrument test between the links. In the first test we established a value for an individual winding and arrived at a value of 1.33Ω, when we then connected the windings in star, the windings are connected in series so the value was double that of the first test $1.33\Omega + 1.33\Omega = 2.66\Omega$.

Continued on page 18.

3 Phase Cage induction motor

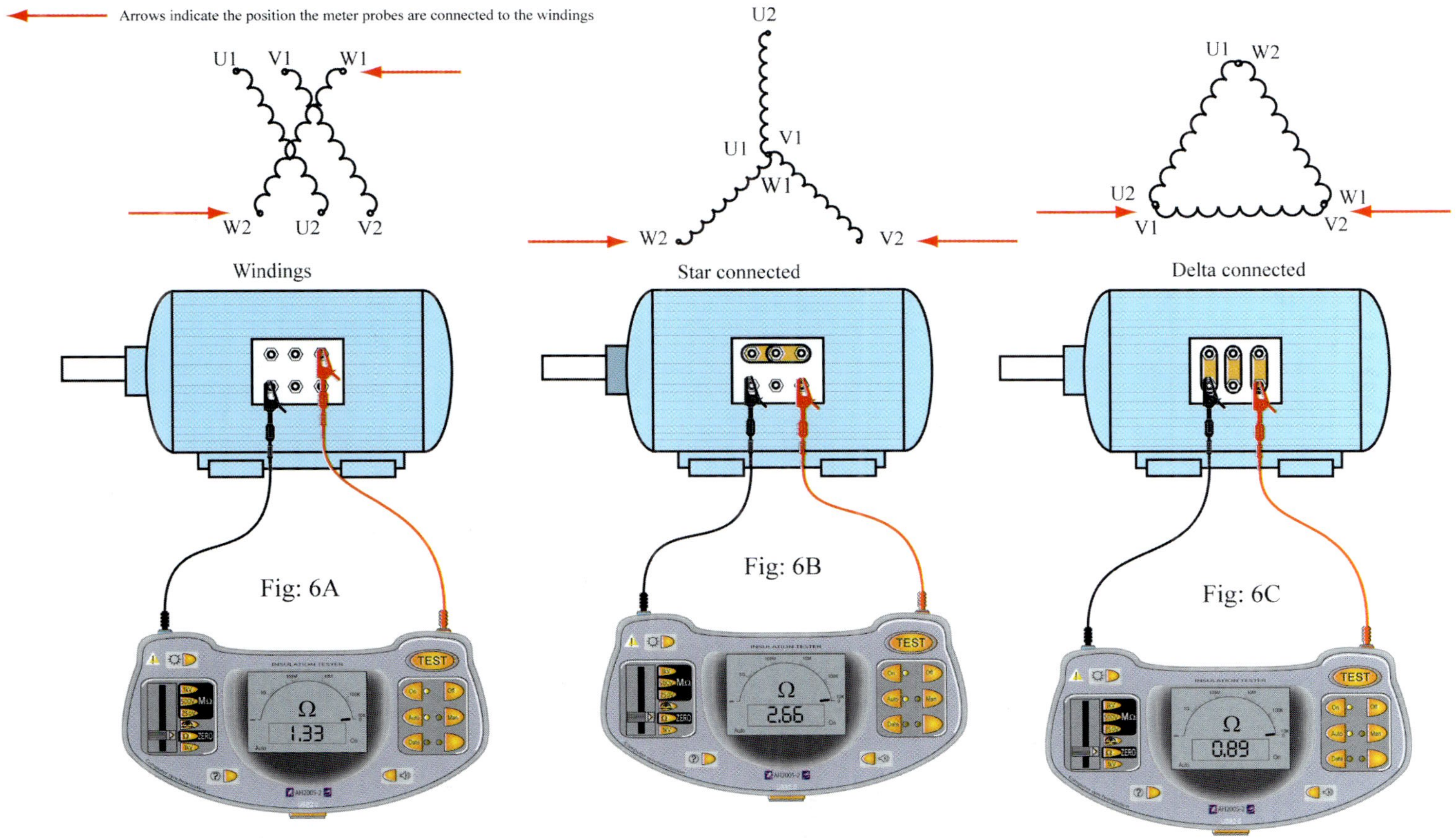

Arrows indicate the position the meter probes are connected to the windings

U1 W1 V1 V2 U2 W2

Windings

Fig: 6A

Ω
1.33

U2 V1 W1 V2 W2 U1

Star connected

Fig: 6B

W1 V2 U1 W2 U2 V1

Delta connected

Fig: 6C

Ω
0.89

Ω
2.55

Fig: 6

The values displayed on the instruments are only examples used to explain the test

17

When the links are in the vertical position as shown in Fig: 6C, we effectively placed two of the winding in series and in parallel with the third winding as shown in Drawing.001.

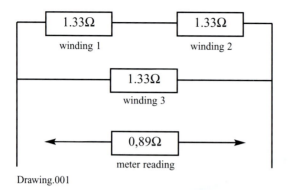

Drawing.001

Let us look at the maths involved in working out what the reading should be.

Firstly add winding 1 to winding 2 1.33Ω + 1.33Ω =2.66Ω.

Now add this value to winding 3, but winding 3 is in parallel so we need to use the reciprocal.

$$\frac{1}{2.66} + \frac{1}{1.33} = \frac{1}{1.128} \therefore = 0,89\Omega$$

Using a scientific calculator the equation goes as follows:

$$2.66 \frac{1}{x} + 1.33 \frac{1}{x} = 1.128 \frac{1}{x} \ 0,89$$

If any of the values are not what you would expect there may be something wrong electrically with the motor. It could be that the links or terminations in the terminal box are loose or dirty, or it could mean that the motor at some stage in it's life has got hot and damaged the windings.

You should carry out insulation resistance as depicted in Fig: 7 on page 19, and as detailed in the section "What is insulation testing?"
Once you are satisfied that all the tests carried out are correct, re-connect the motor cables and connections and replace any gaskets as necessary.

Insulation resistance test on a 3 phase cage induction motor

Set insulation tester to 500V and test

U1 to V1
U1 to W1
V1 to W1
U1 to case of motor or cpc if connected
V1 to case of motor or cpc if connected
W1 to case of motor or cpc if connected
If your readings are not greater than 2MΩ there is something wrong
with the insulation on the windings.

In some cases the tests may show a low insulation resistance
value, this may be due to damaged insulation on the windings,
or moisture on the windings or even infestation by insects.
If you are in any doubt about the serviceability of the motor
you should have it rewound.

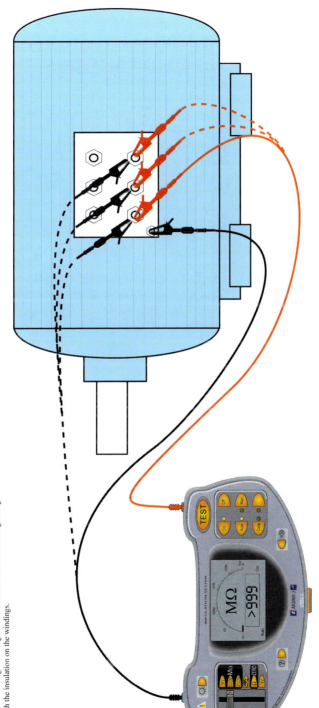

Fig: 7

The values displayed on the instruments are only examples used to explain the test

19

TESTING SINGLE PHASE CAPACITOR START INDUCTION MOTORS

After isolating the motor you must discharge the capacitor before disconnecting the terminals, otherwise you could receive a shock. (see note 8 page 79)

Firstly make yourself familiar with the motor terminations and before you disconnect any cables or terminals make a sketch of how the wiring is connected, including writing down the terminal numbers and how the links are connected, apply the correct labelling if necessary.

Single-phase motors need additional windings and/or components to aid starting, so the wiring is very different to that of a 3-phase motor.

Some single-phase induction motors carry the numbering system for the terminals as indicated in Fig: 8 on page 21, but they do vary depending on manufacturer.

The circuit diagram shows how the motor and supply cables are connected to the terminals in the terminal box.

Sequence for testing:
- Remove all loads
- Ensure Isolation
- Discharge the capacitor by shorting the leads with a test lamp or high wattage lamp with suitable probes (see page 79 note 8)
- Remove capacitor and supply cables from the motor, connect capacitor leads as shown in Fig: 8A
- Select an Insulation continuity tester or multi-meter, set to Ω's and zero
- Connect your meter to the motor across U1 and U2 as indicated in Fig: 8. Press the test button and record the value, if no value is present change the scale until a reasonable value can be recorded.

Refer to the circuit diagram to clarify what test you are carrying out, you have used your instrument to record the value of resistance across terminals U1 and U2 (i.e. 31Ω) this is the resistance of the run winding.

Continued on page 22.

Single Phase Cage induction motor capacitor start

Test: Resistance of the run winding

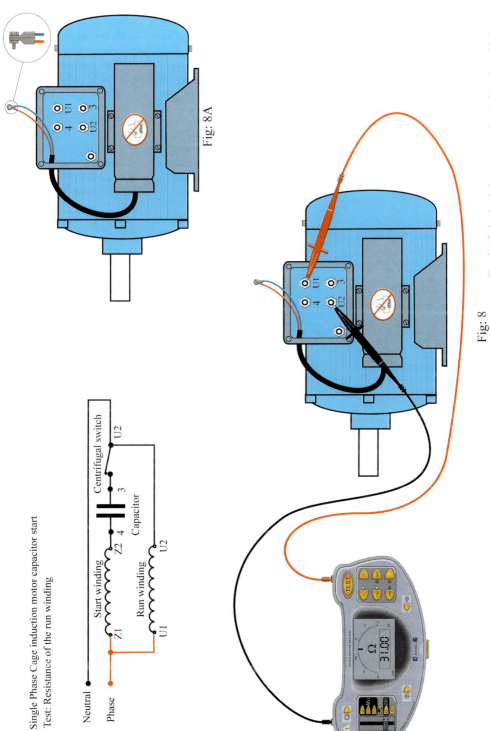

Fig: 8A

Fig: 8

21

Now connect your meter to the motor across terminals 4 and U1 as shown in Fig: 9A page 23, press the test button and record the value, if no value is present change the scale until a reasonable value can be recorded.

Refer to the circuit diagram to clarify what test you are carrying out, you have used your instrument to record the value of resistance across terminals 4 and U1 (i.e. 42Ω) this is the resistance of the start winding.

There are no definitive readings to confirm that the resistance is correct, but the resistance of the start winding must be a greater value than the resistance in the run winding, this has been confirmed.

Run winding = 31Ω

Start winding = 42Ω

Now connect your meter to the motor across terminals 3 and U2 as shown in Fig: 9B page 23, press the test button and record the value, the value should be extremely low 0.01Ω or 0.0Ω.

Refer to the circuit diagram to clarify what test you are carrying out, you have used your instrument to record the value of resistance across terminals 3 and U2 (i.e. 0.01Ω) this is the resistance of the centrifugal switch, which should be negligible.

If you record a high resistance it may be due to an open centrifugal switch or the switch contacts are dirty. (Dirty contacts can be caused by arcing)

How the motor functions
On initial start-up the centrifugal switch is closed and both windings are energised, when the motor reaches a set speed the centrifugal switch opens thus disconnecting the start winding.

If the motor does not run when switched on but makes a buzzing noise it could be that the start winding, run winding or the centrifugal switch is open circuit.

To reverse the direction of rotation on this type of motor you can interchange Z1 and Z2, the start winding or U1 and U2 the run winding.

Notes: Some motors only give access to one winding, some motors only have three exposed terminals and older motors have ABC numbering on the terminals.

Single Phase Cage induction motor capacitor start

Test: Resistance of the start winding and confirmation that the centrifugal switch is closed

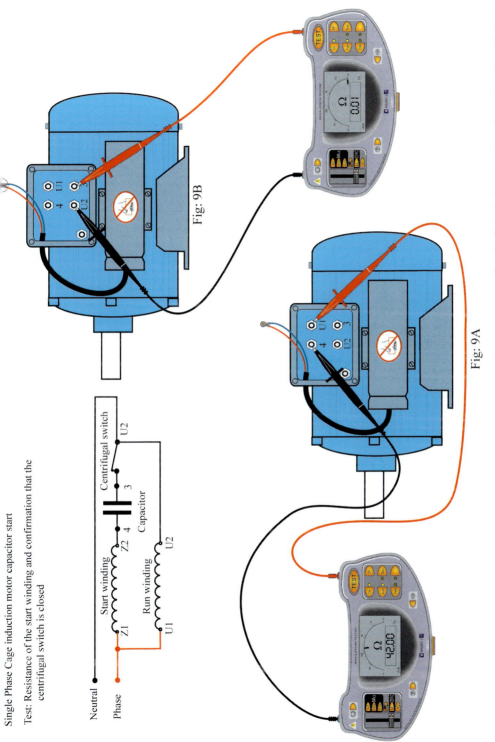

Fig: 9B

Fig: 9A

23

Insulation test:

It is important that an insulation resistance test is carried out between the windings and from the windings to earth, to make sure that the insulation has not broken down.

It is important that you disconnect one of the windings from the terminals otherwise you will only measure the resistance of the windings, and when the instrument is set to the 500v or MΩ range, it may give the impression that the motor is short circuit.

You must be aware that there may be electronic devices in the motor, such as thermal cutouts including thermisters.

Example: If the meter is placed on terminals U2 and 4, the only components in the circuit are the start and run windings, the meter will signify a low insulation resistance. (refer to Fig: 10 page 25 whilst carrying out these tests)

The following sequence should be followed:

> • Set your insulation tester to 500V or the MΩ scale
> • Place the instrument on terminals U1 and earth or cpc if still connected
> • Test and record (value should not be less than 0.5 MΩ)
> • Disconnect Z1 from the terminal box
> • Place the instrument on terminals U1 and Z1
> • Test and record (value should not be less than 0.5 MΩ)

Whilst you are making the tests it would be helpful if you follow the circuit diagram, this will help you determine what components you are testing.

When the instrument was placed on U1 and earth both windings and the centrifugal switch were tested to earth.

When the instrument was placed on terminal U1 and cable Z1, the insulation between the windings was checked, to verify if the insulation had broken down.

Note: Do not use the insulation tester on the capacitor this will only charge the capacitor and give rise to the risk of shock.

Once you are satisfied that all the tests carried out are correct, re-connect the motor cables and connections and replace any gaskets as necessary.

Single Phase Cage induction motor capacitor start

Test: Insulation resistance

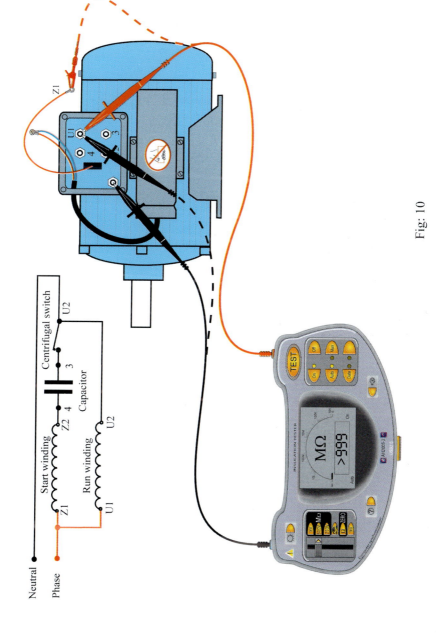

Fig: 10

25

SINGLE PHASE MOTOR OPERATION AND CIRCUIT CONFIGURATIONS

In Fig: 11 page 27 there are various types of common circuit diagrams for single-phase motors.

Basically all single-phase motors operate in similar ways, but completely different to 3 phase motors.

If you look at the sine waves for 3-phase voltages you can see that they are 120° apart, put in a simple form, the brown phase (L1) is energised before the black phase (L2), thus as the windings are energised when connected to separate phases of a 3-phase supply each winding magnetises in turn to create a rotating magnetic field.

The rotating magnetic flux in the stator causes a torque reaction in the rotor by means of electro-magnetic induction and creates a circulating magnetic flux which in turn makes the rotor turn.

If single-phase motors had only one winding a rotating magnetic flux could not be established, so there are two windings, but these windings have to be separated so that a rotating magnetic flux can be established.

There are two ways in which this is done. Firstly in circuit 1 the windings are physically wired in different positions around the stator as shown in the drawing for single-phase currents.

The other method is shown in circuit 2 and 3 where a capacitor is used to electrically shift the winding. (to delay the second winding)

Generally the second winding in single phase motors is shifted by 90 degrees, so that on initial start up the rotating magnetic flux can be established, once the motor reaches a certain speed the start winding can be switched out by a centrifugal switch and therefore leaving the run winding to maintain the rotation of the rotor.

Sine wave for 3 - phase voltages

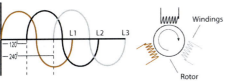

Sine wave for single phase voltage

Single phase induction motor circuit diagrams

Circuit 1] Single phase split phase induction motor

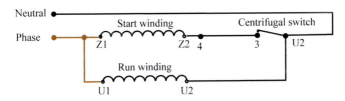

Circuit 2] Single phase capacitor start induction motor

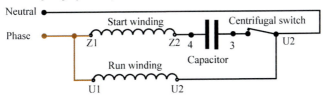

Circuit 3] Single phase capacitor start capacitor run induction motor

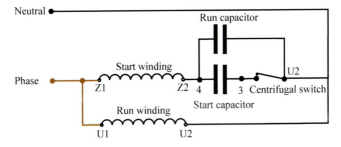

Fig: 11

DIRECT ON LINE & STAR DELTA STARTERS

In Fig:12 page 29 there are various types of motor starter configurations.

It is extremely important that the correct starter is selected for a particular application.

Motor starters have two basic forms either manual or automatic. Manual starters are less versatile and are normally only suitable for infrequent starting of smaller motors.

Automatic starters are rated for frequent duty, high mechanical durability and electrical life, with the facility for remote control.

The most common form of starting a three-phase squirrel cage motor is through Direct on Line and Star-Delta.

The starting current of a standard squirrel cage induction when switched directly on to the supply can be approximately 6 to 8 times the full load current and may develop 150% torque.

However, the magnetic flux produced in the stator of an induction motor rotates immediately the supply is switched on and, therefore, the motor is self-starting.

The motor starter however does not provide starting as the name implies but allows for the motor to be controlled, provide overload and no-volt protection.

BS7671 states that every electric motor having a rating exceeding 0.37 kW shall be provided with control equipment incorporating a means of protection against overload of the motor.

Direct on line starting is generally provided for motors up to 5 kW. The supply authority limit the rating of DOL and above this motors should have star-delta starting, this method of starting restricts the starting current to 2 to 3 times full load current with a corresponding drop in torque.

I have given several examples of the internal control wiring for various starters including three-phase DOL, single-phase DOL, three-phase reversing and star delta.

CPCs have been omitted for clarity

Typical direct-on-line starter
circuit diagrams

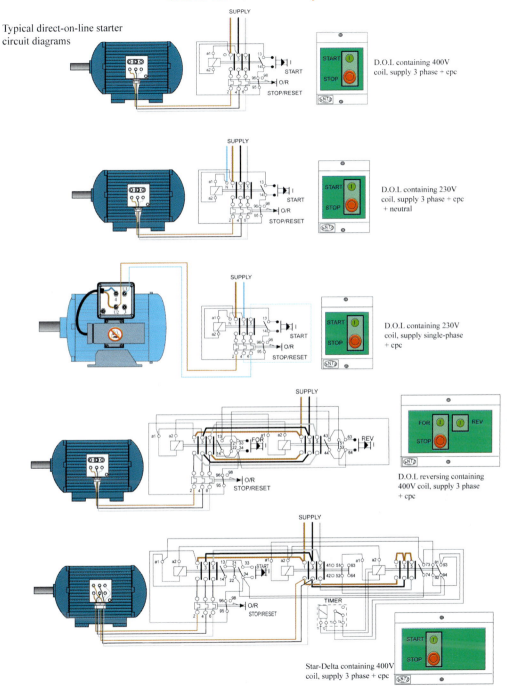

D.O.L containing 400V
coil, supply 3 phase + cpc

D.O.L containing 230V
coil, supply 3 phase + cpc
+ neutral

D.O.L containing 230V
coil, supply single-phase
+ cpc

D.O.L reversing containing
400V coil, supply 3 phase
+ cpc

Star-Delta containing 400V
coil, supply 3 phase + cpc

Fig: 12

UNDERSTANDING CONTROL CIRCUITS IN DOL STARTERS

Before we can look at testing the control circuit in a DOL (Direct On Line) starter it is important that you understand how the circuit functions.

Firstly if you look at the drawing in Fig: 13A page 31, the supply is already supplied from terminal 3 to a1 and from terminal 1 to 14, this is the normal state of the starter.

If we now depress the start button as shown in Fig: 13B the supply is turned on to a2 thus energising the coil and pulling in the contacts.

The supply is now transmitted through the contacts from terminals 1 to 2, 3 to 4 and 5 to 6, thus the motor functions.

When the start button is released as in Fig 13C the starter does not drop out due to the hold on contacts across terminals 13 and 14.

If the motor suffers from an overload either due to mechanical stress or an electrical problem either the overload heaters get hot and bend bimetallic strips in the overloads, or coils sense the excess current causing a piston to rise, both reactions cause the control circuit to open, opening terminals 95 and 96, breaking the supply to the coil.

This in turn drops out the starter and turns off the motor preventing further damage to the motor. (If the overloads are incorrectly set, during fault conditions the motor may be severely damaged)

The overloads can be re-set after the fault has been cleared by pressing the stop/reset button. (Bimetallic strips will need to cool before re-setting can take place)

It is important that you understand how starter circuits function, because there may be a time when you will come across starters without a familiar numbering system, study the drawings and try to follow the control circuit, it will help you later on in your career.

Control circuit operation
on direct-on-line starters

CPCs have been omitted for clarity

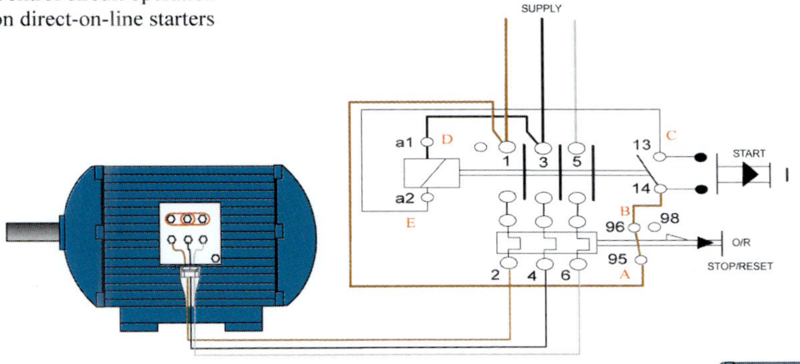

Fig: 13A

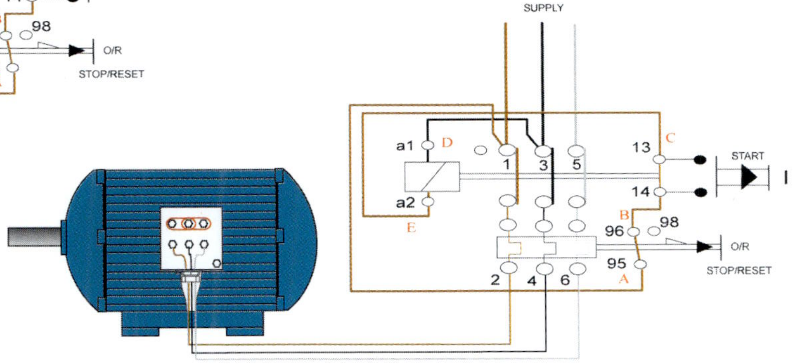

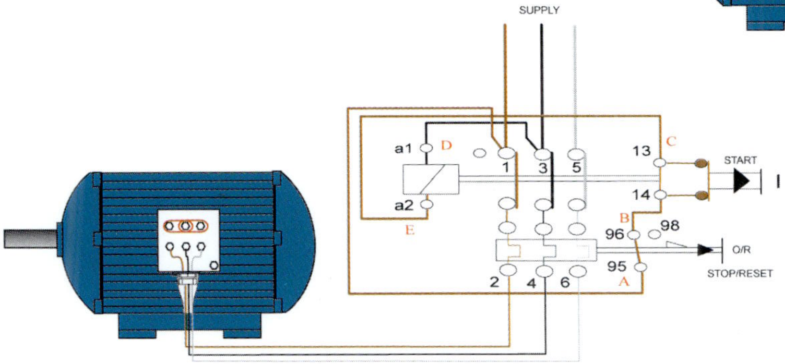

Fig: 13B

Fig: 13C

TESTING THE CONTROL CIRCUIT IN DOL STARTERS

Now that you have an understanding of how the circuit functions we can test the circuit in the dead state.

Follow the circuit diagram in Fig: 14 as the tests are being carried out .

The following sequence should be followed:

Throughout the tests the red probe can stay on terminal 1

1   Carry out isolation procedure (see page 8)
2   Remove any links in the motor terminal box to prevent inaccurate readings
3   Set your insulation tester to the Ω scale and zero instrument
4   From your instrument place the red probe on terminal 1 and the black probe on terminal 95, the instrument should display 0.0Ω, this signifies that the circuit is closed or continuous from terminal 1 to 95 (point A)
5   Now place the black probe on terminal 96, the instrument should display 0.0Ω, this confirms the circuit is continuous from terminal 1 to 96, at this point whilst the meter is still connected, press the stop button, the instrument should change from 0.0Ω to >99,9Ω, you can release and depress the stop button several times to confirm this operation (point B)
6   Place the black probe on terminal 14, the instrument should display 0.0Ω, this signifies that the circuit is now continuous from terminal 1 to 14
7   Now place the black probe on terminal 13, the instrument should display >99,9Ω, as the circuit is now open, this is because we are now across the start button, to maintain continuity, press the start button and the meter should change from >99,9Ω, to 0.0Ω. (point C)
8   At this point in our tests you can also manually depress the starter itself to test the hold on contacts
9   Now place the black probe on terminal a2, the instrument should display 0.0Ω, when the start button is pressed (point E)
10 Now place the black probe on terminal a1, the instrument should display 20.0Ω, when the start button is pressed, the figure of 20.0Ω, is approximate and depends on the type of coil installed, but you should have a resistance, which will confirm that the coil is continuous (point D)

Now that you have confirmed that the control circuit in the DOL starter is ok you can test the main contacts of the starter by placing the test leads across the terminals numbered, 1 & 2, 3 & 4, 5 & 6, and manually pressing in the starter.

Testing the control circuit on direct on line starters

CPC's have been omitted for clarity

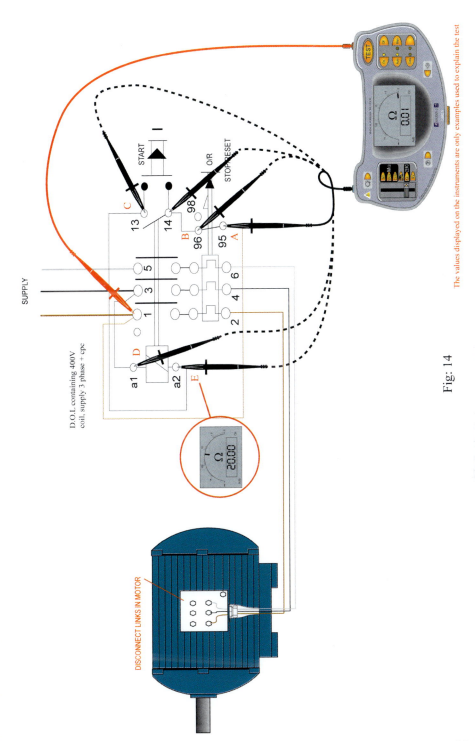

SUPPLY

D.O.L containing 400V coil, supply 3 phase + cpc

DISCONNECT LINKS IN MOTOR

START

O/R

STOP-RESET

The values displayed on the instruments are only examples used to explain the test

Fig: 14

33

MEASURING CURRENT USING CLAMP ON AMMETERS

Current clamp meters have been available for many years now and can measure and record current over time including peak, average and nominal currents using on board recording facilities, they are fairly easy to use, but there are important safety issues that need to be addressed.

Current measurement is necessary to find the actual current drawn by a piece of equipment, and can be useful when diagnosing faults within the equipment.

The type of clamp meter used in the test on page 35 Fig: 15 needs to be clamped directly around the conductor you are trying to measure the current flowing in.

In order to carry out the test you will have to access the enclosures that contain the final circuit cables, and consideration must be given to the isolation procedure every time the meter is connected and disconnected from around the cable.

The final circuit cables have very little protection other than the PVC insulation surrounding the conductors. This insulation is fairly easy to damage, and if you place and remove the instrument from the cable it may damage the insulation, so great care must be taken.

You will also note that the cover has been replaced on the motor terminal box, again this must be carried out very carefully, all terminals and exposed live parts that may be live during the test must be shielded before reinstating the supply to prevent the risk of shock.

Precautions must be taken whilst the test is being carried out as the circuit is live and under load conditions.
The test sequence is as follows:

1 Remove all loads
2 Ensure Isolation
3 Clamp meter on cable and set to the appropriate range
4 Cover other exposed parts that may be made live during the test
5 Reinstate the supply
6 Turn on the equipment and observe reading
7 Turn off supply and ensure isolation
8 Repeat from 3 for other circuit conductors

Remember if you are checking the full load current on motors, they may not be on full load at the time of the test.

Test: Check the current drawn by the motor
using a clamp meter

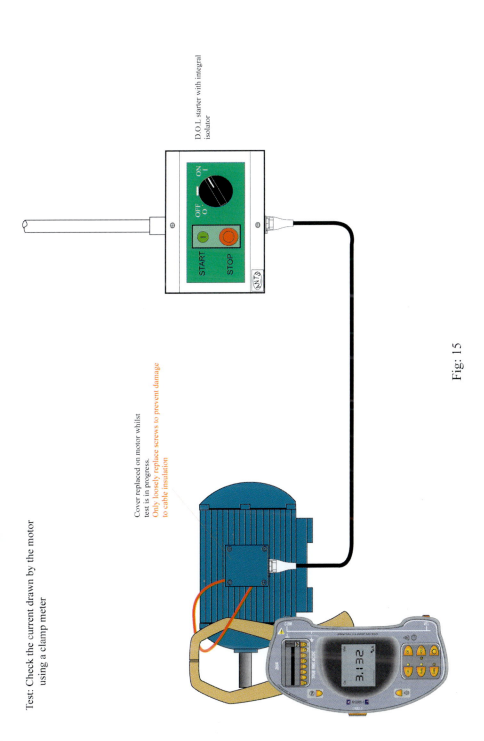

Cover replaced on motor whilst
test is in progress.
Only loosely replace screws to prevent damage
to cable insulation

D.O.L starter with integral
isolator

Fig: 15

The values displayed on the instruments are only examples used to explain the test

35

MEASURING CURRENT USING CLAMP ON AMMETERS

This generation of clamp meters has the ability to measure the current flowing in single-core and multi-core cables with a sheath.

I have shown in Fig: 16 page 37 the meter clamped around the multi-core cable feeding a single-phase motor, I have indicated that the cable is a 3 core flex.

This test can not be carried out on shielded cables so you will need to check if the cable is shielded before the test.

Shielded cables include;

- SWA (Steel Wire Armour)
- MIMS (Mineral Insulated Metal Sheath)
- FP200 (Fire resistant)
- SY flex 3-5 core grey flex with braid and clear over sheath

This instrument is extremely useful for determining the current in T&E (twin and earth), 2 and 3 core flex, as well as any other type of non-shielded cable.

If you are in electrical maintenance this instrument will save you time when diagnosing faults.

The test sequence is as follows:

- Locate equipment to be tested
- Set meter to the appropriate type of cable to be tested
- Set meter to MULTIWIRE or SINGLEWIRE
- Clamp meter on cable and record value

With this instrument you avoid the risk of exposing yourself to live parts, you will save time not having to shield other live parts and you do not need to perform safe isolation.

Whilst using the instrument to diagnose faults you will not have to turn off equipment, so down time is minimised.

Test: Check the current drawn by the motor using a clamp meter

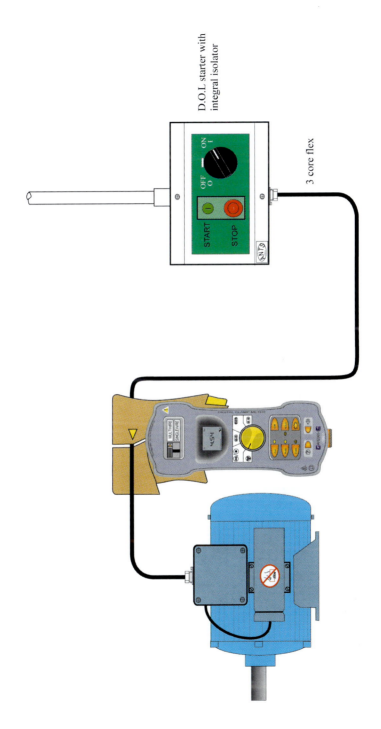

D.O.L. starter with integral isolator

3 core flex

Fig: 16

UNDERSTANDING SOIL RESISTIVITY

Wiring systems TN-C, TN-S and TN-C-S have their earthing provision supplied from the source of energy and all the protective conductors use this earthing provision to create disconnection of the protective device during an earth fault, however, TT and IT systems do not have a supplied earthing source and therefore need to have a means of earthing the wiring system.

This means creating a correctly designed quality earthing system using rods, plates mesh, mats, tapes and any combination of earthing conductors to achieve a low enough resistance path for the earth fault to travel quickly enough to earth to disconnect the protective device.

In order to disconnect a system during an earth fault within a set time we have to consider the resistance that the earthing system offers to the flow of the current along the path to the general mass of earth.

Whilst you would know the resistance of the main protective conductor and any components connected to it, you will not know the resistivity of the soils in which you are to create your earth.

There are several factors that affect soil resistivity, such as physical composition.

Soil type	Resistivity Ω.m	Soil type	Resistivity Ω.m
Marshy	2 – 2.7	Peat	$\geq$200
Loam and clay	4 – 150	Sandy gravel	300 – 500
Chalk	60 – 400	Rock	$\geq$1000
Sand	90 – 8000		

Increased moisture in the ground can rapidly decrease resistance, you must consider areas where there are variations in rainfall throughout the year, chemical composition of the soils as certain minerals and salts when increased reduce resistivity and finally temperature, the colder the ground gets the higher the resistance so getting below the frost line is important.

If you are involved in the design and location of an earthing system or looking for a resistance reading less than 1Ω then you must use a 4 terminal tester as this will give a more precise measurement of the ground resistance only.

Having looked at the resistance of the conductors, the environmental conditions and soil composition, you need to know that an earth electrode does not have a single point of contact with the soil, it is a volumus conical contact dependant on the type and size of the electrode. See fig17 page 39

Understanding earth electrode resistance

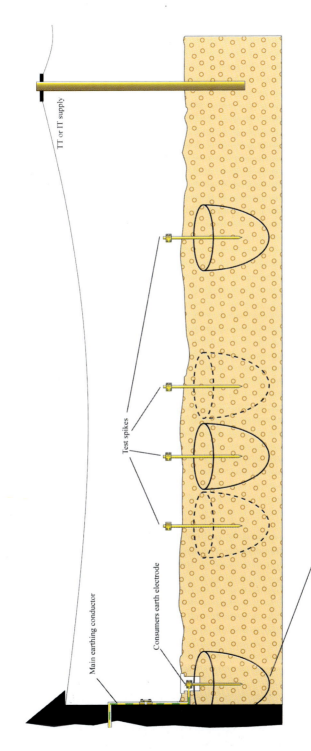

Fig: 17

The values displayed on the instruments are only examples used to explain the test

EARTH ELECTRODE RESISTANCE MEASUREMENT

This simplistic view in Fig: 18 page 41 is an example of carrying out an earth electrode test using a 3 terminal test meter. If using a 4 terminal test meter connect C1 & P1 to the earth electrode (null test), this will exclude the resistance of the test leads from the resistance value, do not short out the terminals C1 and P1, unless the test leads have insignificant resistance.

The distance between the electrode and the current spike C2 is ten times the length of the electrode under test, i.e. if the electrode is 3m long, the current spike will be 30m away, this will produce reliable test results. (It will be difficult to determine the length of the earth electrode)

A potential spike (P2) is placed midway between the electrode and current spike (C2) (point A), then 10% the distance between electrode and current spike (point B), and 10% the distance between electrode and the current spike (point C).
In this case it is placed 3m either side of the current position of the potential spike (P2).

The installation must be shut down and isolated as a precaution against the possibility of a fault developing whilst you are carrying out the tests, and giving rise to the risk of shock and fire.

It is considered good practice to remove the main earthing conductor from the earth electrode and not at the test point or earthing terminal.

Test sequence:
- Shut down the installation and isolate the supply
- Disconnect main earthing conductor at the earth electrode
- Place spikes at locations A and D
- Connect instrument, take reading and record
- Move potential spike to position B, take reading and record
- Move potential spike to position C, take reading and record

Example: Let us assume that the first reading taken with spikes in position A and D was recorded as 50Ω, then in position B and D the reading was 45Ω, and at position C and D was 52Ω. The average value is 49Ω, therefore the percentage difference is (49Ω - 45Ω) = 4Ω, ∴ 4Ω ÷ 49Ω x 100% = 8% differential which is a greater value than the maximum 5% value allowed. In this case a longer or extra electrode will need to be fitted and the tests repeated.

Test: Earth electrode test using earth tester

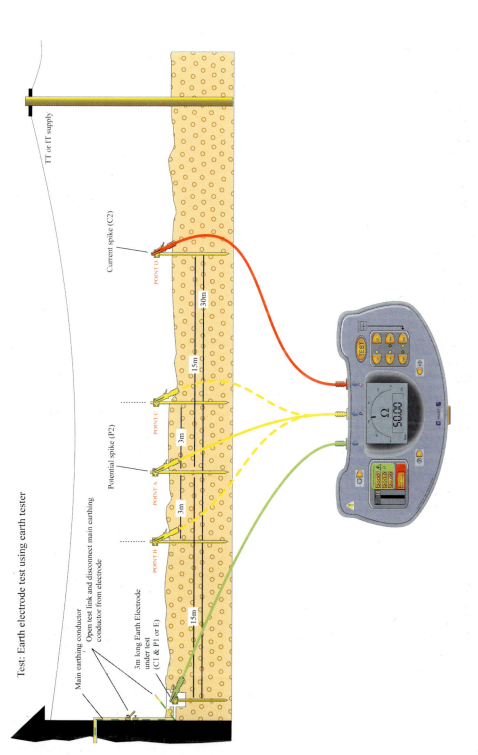

Main earthing conductor

Open test link and disconnect main earthing conductor from electrode

3m long Earth Electrode under test (C1 & P1 or E)

Potential spike (P2)

Current spike (C2)

TT or IT supply

POINT B

POINT A

POINT C

POINT D

15m

3m

3m

15m

30m

Fig: 18

The values displayed on the instruments are only examples used to explain the test

ALTERNATIVE EARTH ELECTRODE RESISTANCE MEASUREMENT USING EARTH LOOP TESTER

The sequence is as follows:
- Remove all loads
- Ensure Isolation
- Disconnect earthing conductor
- Select an earth loop tester and set to largest Ω scale
- If the installation is protected by an RCD select the $200\Omega/2k\Omega$ scale
- Place instrument on the circuit as shown (Phase & Earth)
- Take readings and record

This is classified as **live** testing so **precautions must be taken** to ensure yours and everyone's safety during the tests. To carry out the test you can place the test leads on the supply point or incoming supply or supply side of the main switch, please be aware that you will not be able to isolate this side so all your safety procedures must be in place.

You will need a special lead set conforming to HSE GS38.

Example: You are trying to determine the resistance of the earth electrode. Record the value taken phase to earth, (treat the impedance reading as the electrode resistance ie 0.35Ω) then you add the resistance of the cpc for the final circuit (assume 0.6Ω). The combination of the resistance in Ω and the RCD operating current in amperes must not exceed 50V for normal dry conditions and 25V for construction, agricultural & horticultural sites.

So if we then take the value of the RCD tripping current as 30mA = 0.03A x (0.35Ω + 0.6Ω) = 0.03V. If this value is greater than 50V carry out test on page 40 & 41 to determine the actual resistance of the earth electrode.

Important note: Remember to reconnect the main earth conductor on completion of the tests.

RCD operating current IΔn	Max values of earth electrode resistance, RA	
	During normal dry conditions	Construction, agricultural & horticultural sites
30mA	1660Ω	830Ω
100mA	500Ω	250Ω
300mA	160Ω	80Ω
500mA	100Ω	50Ω

Test: Earth electrode test using earth loop impedance tester

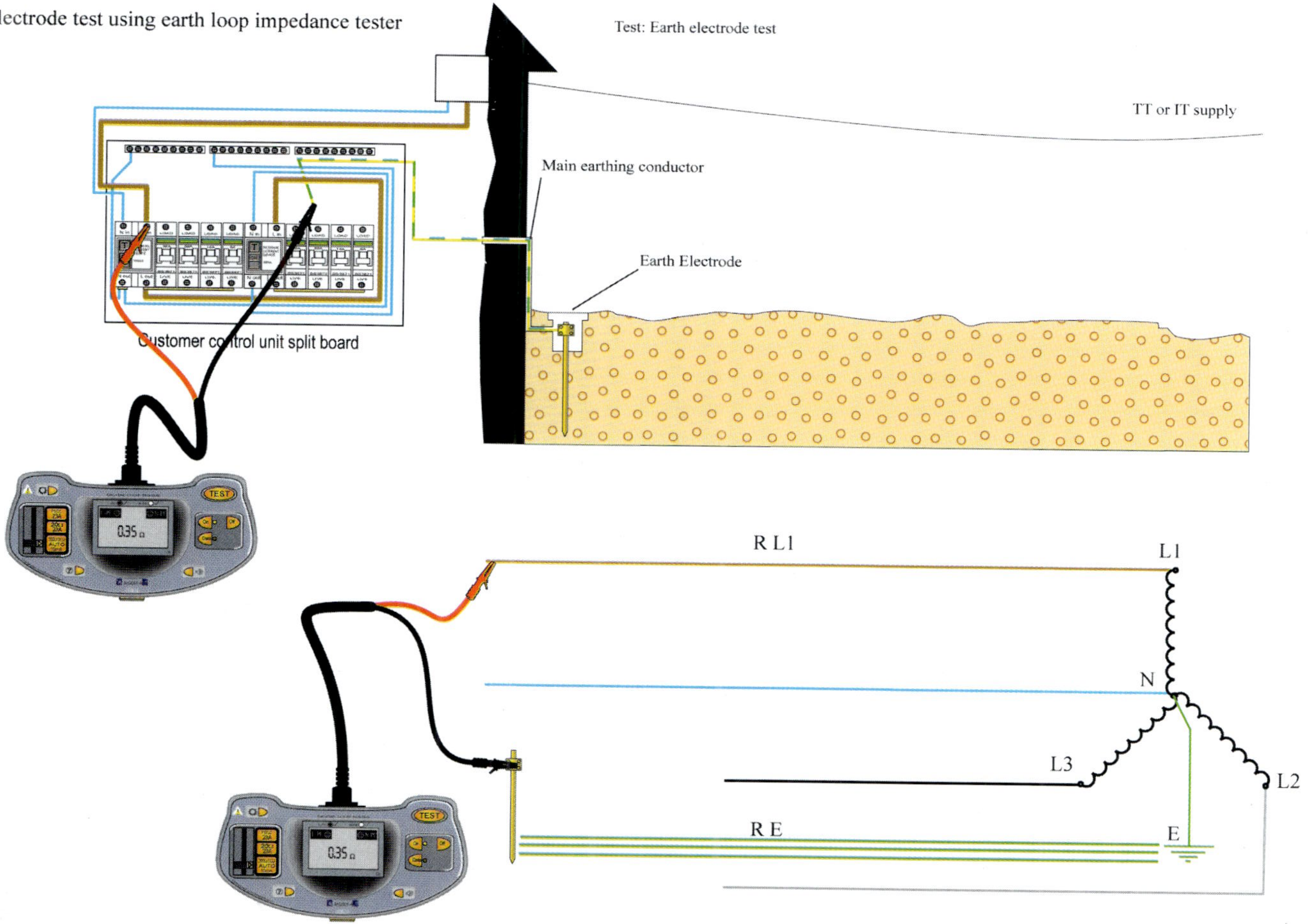

Test: Earth electrode test
TT or IT supply
Main earthing conductor
Earth Electrode
Customer control unit split board
TEST
0.35 Ω
R L1
L1
N
L3
L2
R E
E

Fig: 19

The values displayed on the instruments are only examples used to explain the test

UNDERSTANDING DOMESTIC CENTRAL HEATING SYSTEM 'Y' PLAN

All electrical systems need you to have a basic knowledge of how they function in order for you to wire, maintain and carry out fault finding.

Here I will be looking at how a 'Y' plan domestic heating system functions, and what each component in the system do.

The type of heating system in Fig:20 is termed 'Y' plan because a 3-way valve is used to direct the hot water around the system.

No heating system should function without some form of automatic control, modern systems are designed for optimum efficiency, reliability, and to provide a comfortable environment in which to live.

The heart of any heating system is its boiler which can be fuelled by mains gas, low pressure gas or oil, and can be quite complicated, so I will only concentrate on the electrical control systems external to the boiler.

The system shown in Fig:20 is a fully pumped system, this means that the water in the system is circulated around the whole of the system by a pump.

A combination of time clock/programmer, room thermostat and hot water thermostat are used to control the system informing the boiler and it's associated equipment whether hot water and/or heating are required and when.

The programmer switches on the boiler and pump so that the circulating water in the system can be heated, this hot water is then sent to the 3-way valve which directs the hot water either to the hot water cylinder or the radiators or both depending on the temperatures of the space to be heated and the water in the hot water cylinder.

If the water in the cylinder is cold the hot water thermostat tells the 3-way to direct the water through the cylinder to heat the domestic hot water, and if the room temperature is cold the room thermostat tells the valve to direct the water through the radiators to heat the rooms.

If however both thermostats detect that the temperature in the hot water cylinder and the rooms is already at the temperature required, the boiler is switched off until the temperature drops sufficiently enough to bring the boiler back on.

Domestic central heating system 'Y' plan

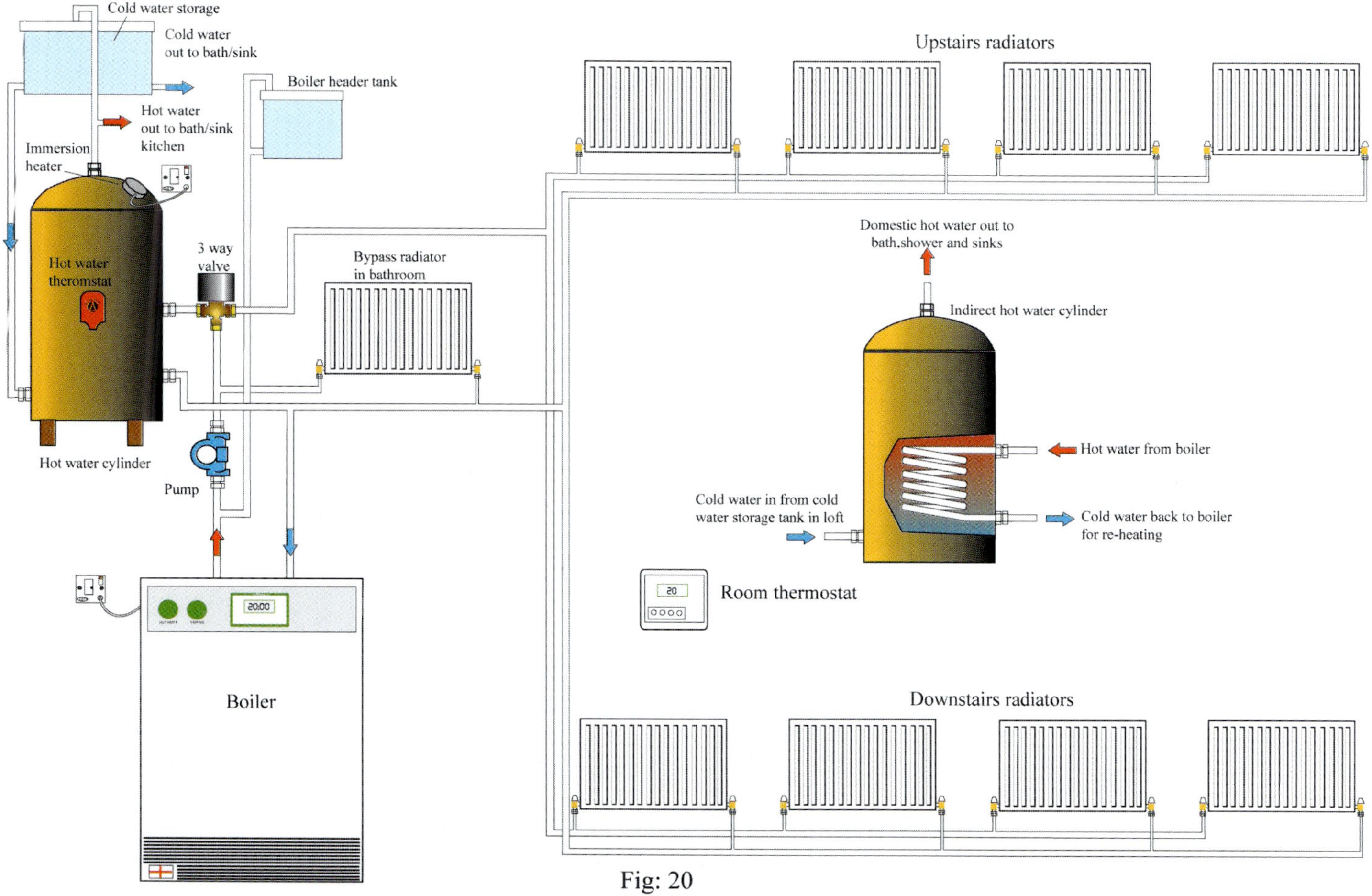

Upstairs radiators

Downstairs radiators

Domestic hot water out to bath,shower and sinks

Indirect hot water cylinder

Hot water from boiler

Cold water back to boiler for re-heating

Cold water in from cold water storage tank in loft

Room thermostat

Boiler header tank

Cold water storage

Cold water out to bath/sink

Hot water out to bath/sink kitchen

Immersion heater

Hot water theromstat

Hot water cylinder

3 way valve

Bypass radiator in bathroom

Pump

Boiler

Fig: 20

45

UNDERSTANDING DOMESTIC CENTRAL HEATING SYSTEM 'S' PLAN

The type of heating system in Fig:21 is termed 'S' plan because two 2-way valves are used to direct the hot water around the system.

No heating system should function without some form of automatic control, modern systems are designed for optimum efficiency, reliability, and to provide a comfortable environment in which to live.

The system shown in Fig:21 is a fully pumped system, this means that the water in the system is circulated around the system by a pump similar to that in Fig: 20 page 45.

A combination of time clock/programmer, room thermostat and hot water thermostat are used to control the system telling the boiler and it's associated equipment whether hot water and/or heating are required and when.

The programmer switches on the boiler and pump so that the circulating water in the system can be heated, this hot water is then sent to the two 2-way valves which directs the hot water either to the hot water cylinder or to the radiators or both depending on the temperatures of the space to be heated and the water in the hot water cylinder.

If the water in the cylinder is cold the hot water thermostat signals the hot water 2-way valve to direct the water through the cylinder to heat the domestic hot water for washing, and if the room temperature is cold, the room thermostat signals the central heating 2-way valve to direct the water through the radiators to heat the rooms.

If however both thermostats detect that the temperature in the hot water cylinder and the rooms is already at the temperature required, the boiler is switched off until the temperature drops sufficiently enough to bring the boiler back on.

Both 'Y' and 'S' plan systems function in very similar ways but with one significant difference, there are two 2-way valves in the 'S' system and a 3-way valve in the 'Y' system. 'Y' plan systems are more common in domestic installations because one 3-way valve is cheaper to purchase than two 2-way valves, but if you want better control over the heating system 2-way valves can be fitted to the upstairs and downstairs radiators independently, and therefore if a second room thermostat was added upstairs the system would be more flexible and more efficient.

Domestic central heating system 'S' plan

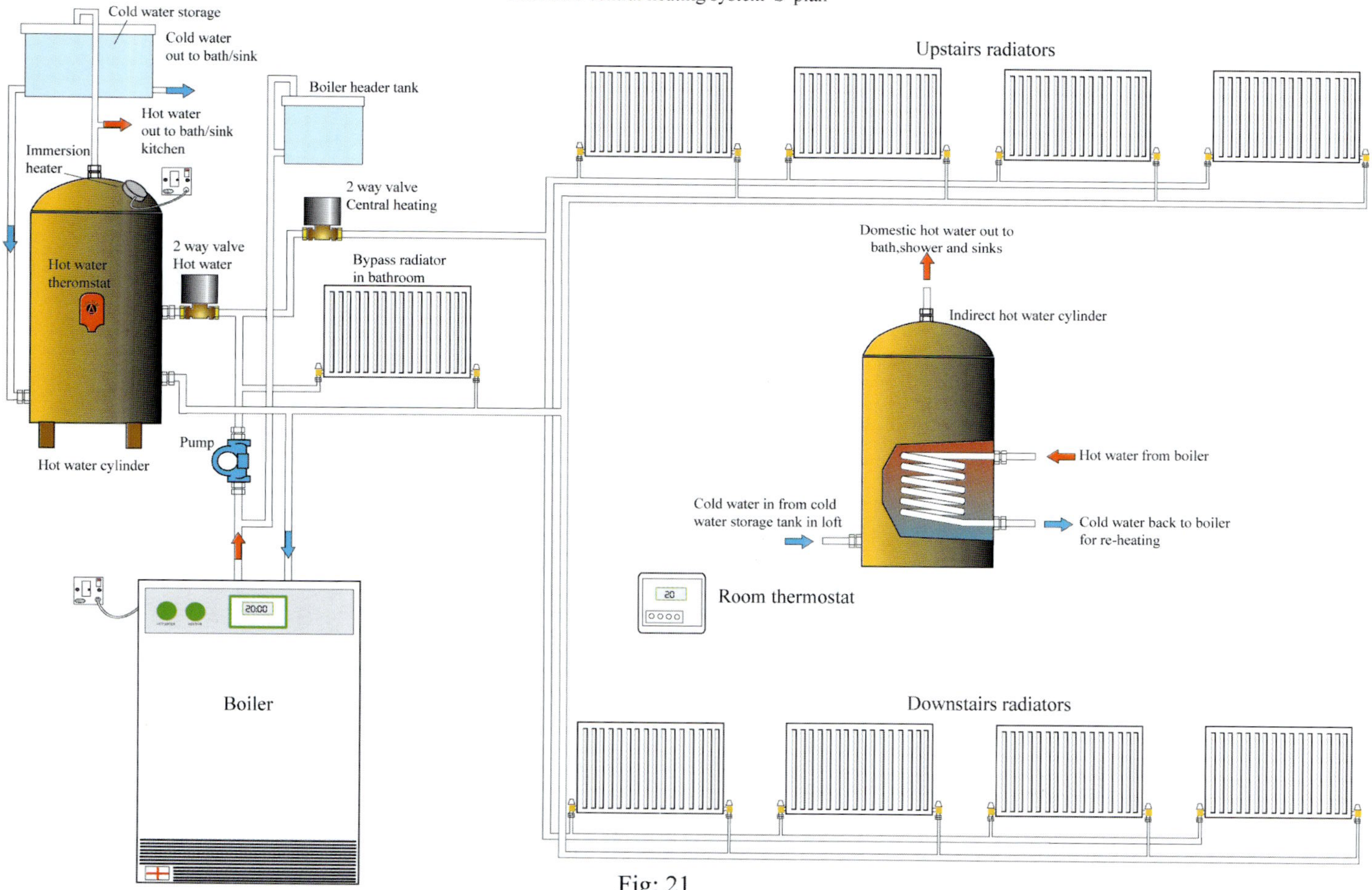

Upstairs radiators

Downstairs radiators

Hot water from boiler

Cold water back to boiler for re-heating

Indirect hot water cylinder

Domestic hot water out to bath, shower and sinks

Cold water in from cold water storage tank in loft

Room thermostat

Fig: 21

Boiler header tank

2 way valve Central heating

Bypass radiator in bathroom

Cold water storage

Cold water out to bath/sink

Hot water out to bath/sink kitchen

2 way valve Hot water

Immersion heater

Hot water theromstat

Hot water cylinder

Pump

Boiler

TESTING THE COMPONENTS WITHIN A DOMESTIC 'Y' & 'S' PLAN HEATING SYSTEM

Even though the wiring diagrams on page 49 and 51 at first seem a little complex, they are fairly straight forward when it comes to testing and fault finding.

All you have to do is split the circuit into it's individual components and test each individual item separately, once all of the components have been tested you can then test the conductors for continuity.

From experience I have found that it is generally unlikely that the circuit conductors break between two points but not un-heard of, so start by testing the components first and then the fixed wiring.

For these tests you will need to refer to the drawings and the text on pages 48 to 51, as I felt that it was not necessary to repeat some tests as they are common to both circuits.

If the object of testing the circuits is to confirm the correct function of the components, or to find a faulty piece of equipment, the starting point must be to confirm that there is a live supply feeding the system and that any protective device (fuse, overload, over-temperature, etc) is intact.

You do not want to spend hours stripping down and testing all of the wiring and its components before you find that the fault was caused by a blown fuse which may have directed you to the fault more quickly.

After you have ruled out any supply failure carry out safe isolation and connect your continuity test meter as shown at position A. You are only testing the open and closed contacts on the room thermostat, and by increasing and decreasing the temperature on the thermostat you can simulate the switching on and off.

With the meter connected at position B you can see that the cylinder thermostat is tested in the same way as the room thermostat.

When testing the pump make sure that any capacitors have been discharged, then connect your meter as shown at position C, you should expect to read a resistance value to show that the windings are continuous and not open circuit or short circuit.
Position E & F are however a little different, and you will need to study the components in the 3-way and 2-way valves.

Test: Check wiring and equipment on 'Y' plan heating systems

CPCs have been omitted for clarity

The values displayed on the instruments are only examples used to explain the test

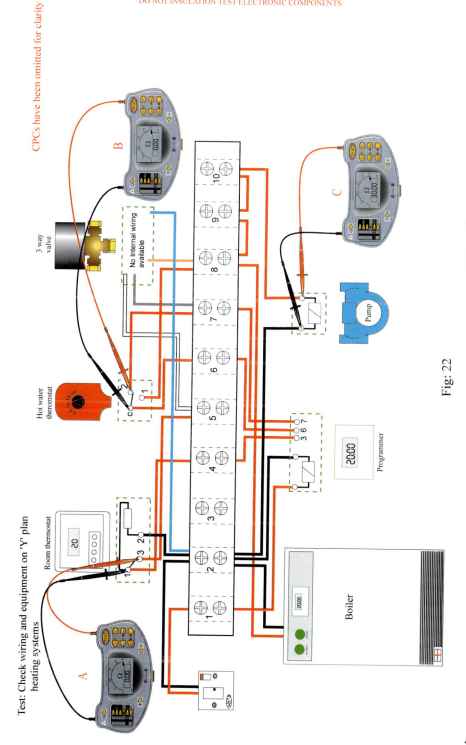

3 way valve

No Internal wiring available

Hot water theromstat

Room thermostat

Programmer

Boiler

Pump

Fig: 22

Both of these valves have manual over-rides so that in the event of failure they can be locked in the open position, you can use this facility to simulate the opening and closing of the valves.

Firstly test the coil on the small motor that makes the valve function. The reading on the meter should show a resistance, signifying the coil is continuous, secondly using the over-ride you can test the micro switches inside the valve for open and closed contacts.

Because of the many different varieties of programmers and time-clocks it is virtually impossible to cover the testing of these devices and, in the long run may cost more to fault find that it would if you were to replace them.

Over the years programmers have become throw-away items, and you may find that a lot of the electronic components now used for controlling electrical circuits are sealed and access is difficult without destroying the unit.

But, if you have tested all of the other components that make up the circuit and eliminated the wiring then the only component left would be the programmer/time-clock.

Testing the wiring is a matter of following the circuit diagram, and tracing the cables as they follow the circuit, to confirm continuity, all you are checking is to make sure that the wiring goes from one point to the other, according to the wiring diagram and that there are no breaks.

I have indicated with the test at position D, how this is carried out, firstly connect one lead of the meter to the point on the cable you wish to test and proceed around the circuit until either you confirm that it is an open or closed circuit.

As we are dealing with continuous conductors the reading on your meter should always signify a very low or negligible resistance.

It may also be necessary to use long test leads when testing, for instance the boiler may be in the kitchen and the terminal box in the cupboard with the hot water cylinder. You will need to use your expertise in deciding how you are going to physically carry out these tests.

Remember no two systems are exactly the same, and only experience can eventually give you all the necessary skills you will need.
All I have done is help you think along the lines of splitting the circuits into small individual components that can be tested.

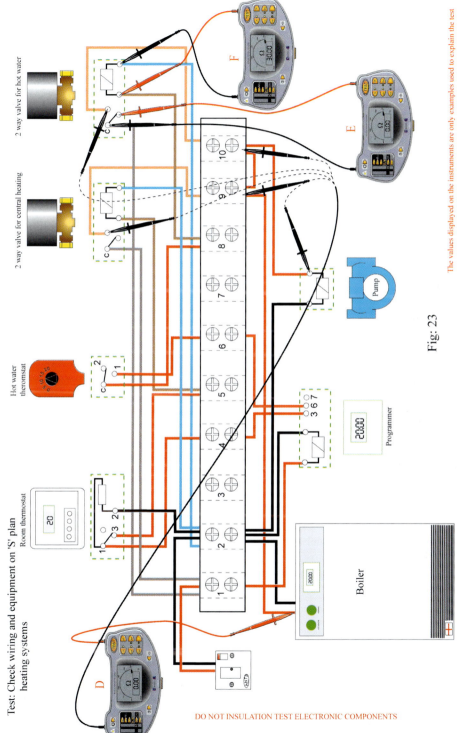

CPC's have been omitted for clarity

Test: Check wiring and equipment on 'S' plan heating systems

2 way valve for hot water

2 way valve for central heating

Hot water thermostat

Room thermostat

Pump

Programmer

Boiler

Fig: 23

The values displayed on the instruments are only examples used to explain the test

DO NOT INSULATION TEST ELECTRONIC COMPONENTS

RAMP TESTING RCDs

Ramp testing is not currently a requirement or a necessity, so what is ramp testing and why carry out the test?

As electricians you only test RCDs to make sure that they do not trip when applying 50% of the RCD rated tripping current, and you record the disconnection time in milli seconds for a 100% of the RCD rated tripping current.

However there is an ever-increasing need to verify the actual magnitude of the fault current that disengages the trip, this can be recorded using the ramp test.

Basically after the instrument has applied a 50% fault current of the rated value of the trip, in which the trip should not disengage, it then performs the ramp test which increases the fault current in increments (indicated in Table.001 below) at 200 millisecond intervals until the RCD trips, the current at which the trip disengages is then displayed on the meter.

The magnitude of the fault current increments is tabulated below and is dependent on the RCD nominal trip value.

RCD nominal trip value	Current range	Step value
10mA	5.. 15mA	1mA
30mA	15.. 50mA	1mA
100mA	50.. 150mA	5mA
300mA	150.. 300mA	5mA
500mA	250.. 500mA	10mA
1000mA	500.. 1000mA	20mA

Table.001

So what do you do with the test result values?

Well, if you were diagnosing faults on RCDs, an oversensitive trip will trip at too low a fault current and may cause nuisance tripping.

An insensitive trip will trip at too high a fault current and may prevent timely disconnection in a fault situation, that is to say that neither of these fault situations may be diagnosed with current test methods.

Depicted in Fig:24 on page 53 and Fig:25 page 55 are examples of carrying out ramp testing on a ring final circuit and a lighting circuit. (The corresponding text is on this page)

Ring final circuit
Test: Residual current ramp test

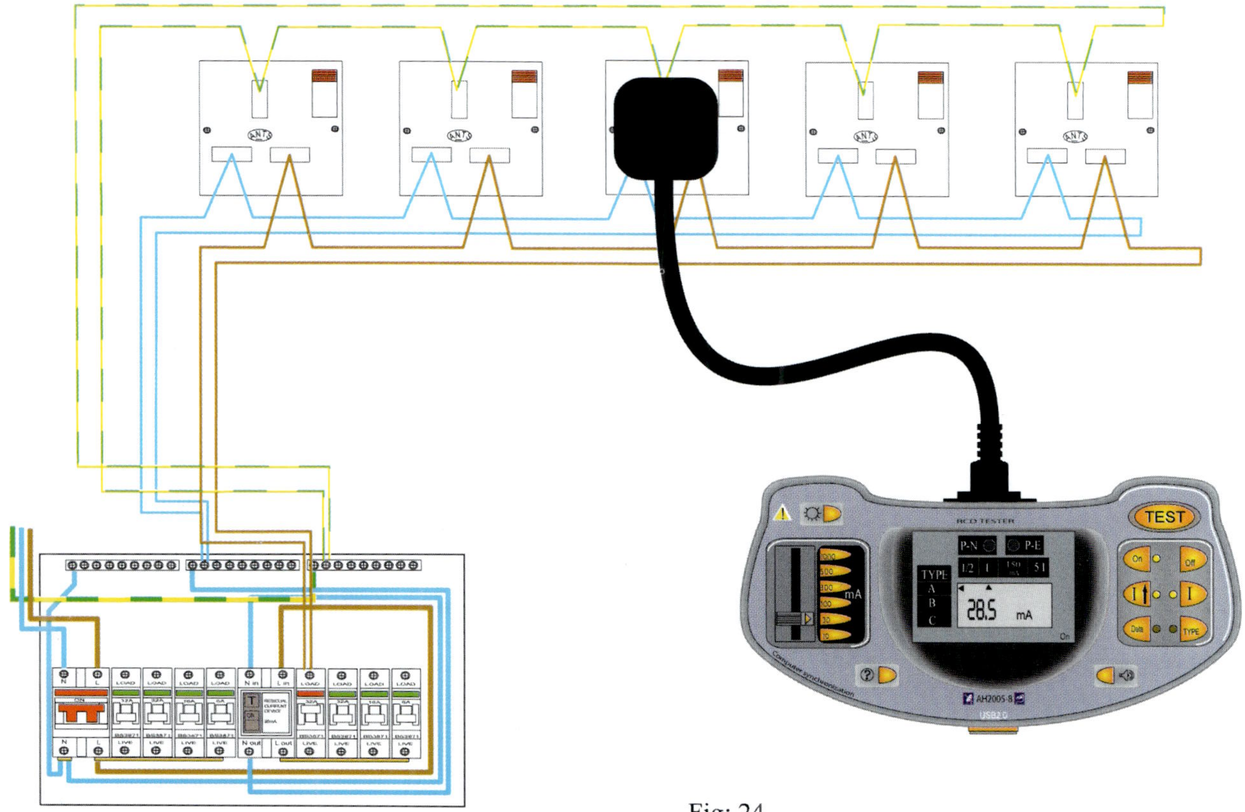

Fig: 24

Customer control unit split board

RAMP TESTING RCDs IN RING FINAL & LIGHTING CIRCUITS

The sequence is as follows:

- Remove all loads
- Ensure Isolation
- Select the appropriate RCD rated current
- Connect to the remote points in the circuits as indicated in Fig: 24 & 25
- Turn on supply
- Select ramp test using the **'I'** key
- Select the RCD type using the **'TYPE'** key
- Press and hold **'TEST'** button
- If the RCD trips, the contact or fault voltage is displayed. The loop or earth resistance and trip current can be displayed by pressing the **'DISPLAY'** key

Remember that these tests are classified under the Health and Safety as live testing so **precautions must be taken** to ensure yours and everyone's safety during the tests.

There is no excuse for dangerous practices; you are personally liable for your actions.

This test must be carried out at the remote point in the circuit, the remote point in a ring final circuit is the centre socket on the ring. You may have already confirmed this when establishing confirmation of the ring through continuity tests.

In the case of lighting and other radial circuits you will need to connect the meter at the last outlet where the final load is connected to the circuit.

Note: In table.001 on page 52 the Current range refers to the starting current and the maximum current, i.e. If you are testing a 30mA device the start of the ramp test is at a value of 50% of 30mA which is 15mA increasing at 200ms intervals in 1mA stages until the device either trips or reaches 50mA.

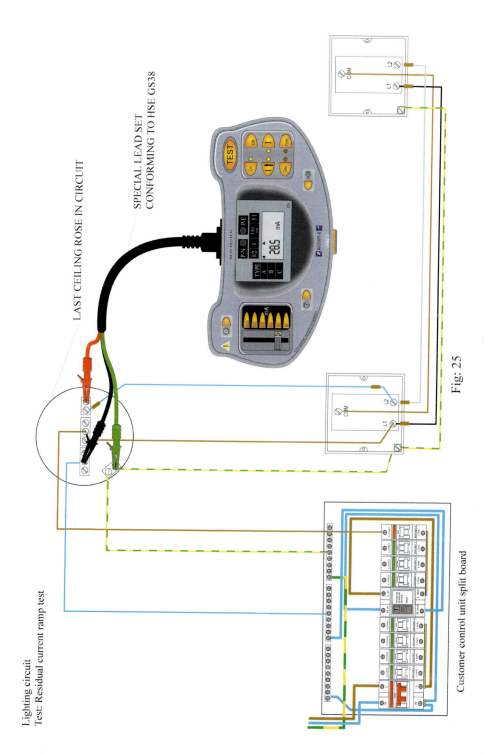

LAST CEILING ROSE IN CIRCUIT

SPECIAL LEAD SET
CONFORMING TO HSE GS38

Fig: 25

Lighting circuit
Test: Residual current ramp test

Customer control unit split board

UNDERSTANDING SWITCH START FLUORESCENT LIGHTING CIRCUITS

Fluorescent fittings are categorised as discharge lights and as I have previously mentioned you need to have an understanding of how the components in the circuit function, and what they are supposed to do, before you can decide on how the circuit and its components can be tested.

There are two fittings depicted in Fig: 26 on page 57. Fig: 26A is a single tube fitting and Fig: 26B a twin fitting.

Both of these fittings are termed as switch-start, they incorporate a glow type starter which consists of two bi-metallic strip electrodes encased in a glass bulb containing an inert gas.

The switch starter is normally open, and when the supply is switched on the full mains voltage is developed across these contacts and a glow discharge takes place between them.

The switch electrodes start to warm and bend towards each other until they meet, when the electrodes meet the glow discharge between the electrodes is extinguished as no voltage is developed across the electrodes, the bi-metallic strips then cool and the electrodes open.

The choke is in series with the lamp and the collapsing of the inductive circuit causes a large enough voltage to strike an arch in the tube, the operation of the circuit is similar to that of the coil and points in a car, the points closing collapses the field in the coil which in turn causes a large voltage across the spark plug which in turn ignites the fuel in the cylinders.

If the tube does not light the whole process is repeated until the tube lights.

Discharge lighting
Basic Fluorescent tube circuit diagrams

Single tube fitting

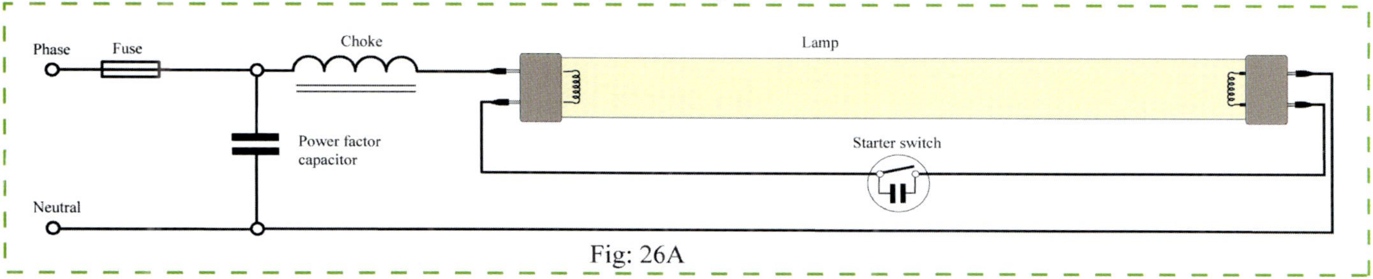

Lamp

Starter switch

Choke

Power factor capacitor

Fuse

Phase

Neutral

Fig: 26A

Double tube fitting

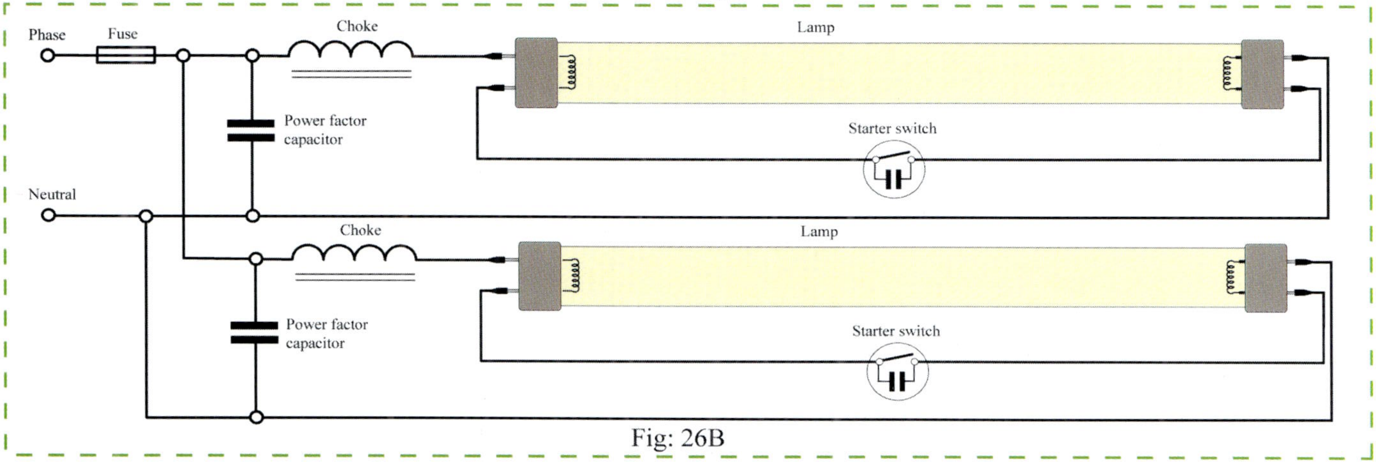

Lamp

Lamp

Starter switch

Starter switch

Choke

Choke

Power factor capacitor

Power factor capacitor

Fuse

Phase

Neutral

Fig: 26B

Fig: 26

TESTING COMPONENTS IN SWITCH START FLUORESCENT LIGHTING CIRCUITS

The sequence is as follows:

1   Remove all loads

2   Ensure Isolation

3   Discharge the capacitor using a test lamp or high wattage lamp, then disconnect the capacitor from the circuit

4   Disconnect fitting from the circuit

5   Remove starter switch and short out terminals

6   Select an insulation continuity tester and set to Ω scale and zero instrument

7   Place the black lead in the Phase conductor terminal in the fitting

8   Place the red lead on the other side of the protective device (fuse if the fitting has one)

9   Take reading and record, the value should be very small typically 0.0Ω

10 Move the red lead to the output side of the choke or the lamp terminal

11 Take reading and record, the resistance should not be 0.0Ω as this signifies a short circuit, there is no absolute reading, you should expect a resistance value

12 Now move the red lead to the other end of the tube, the value should be similar to that recorded in 11 above

13 Now place the red lead on the neutral terminal, and again a similar value should be displayed as in 11 & 12 above

If the resistance reading across the choke is negligible it likely that the choke is short circuit, if continuity is lost at any point it may be the lamp ends and or the wiring within the fitting are open circuit.

You can adapt an old switch starter to take the place of the shorting link.

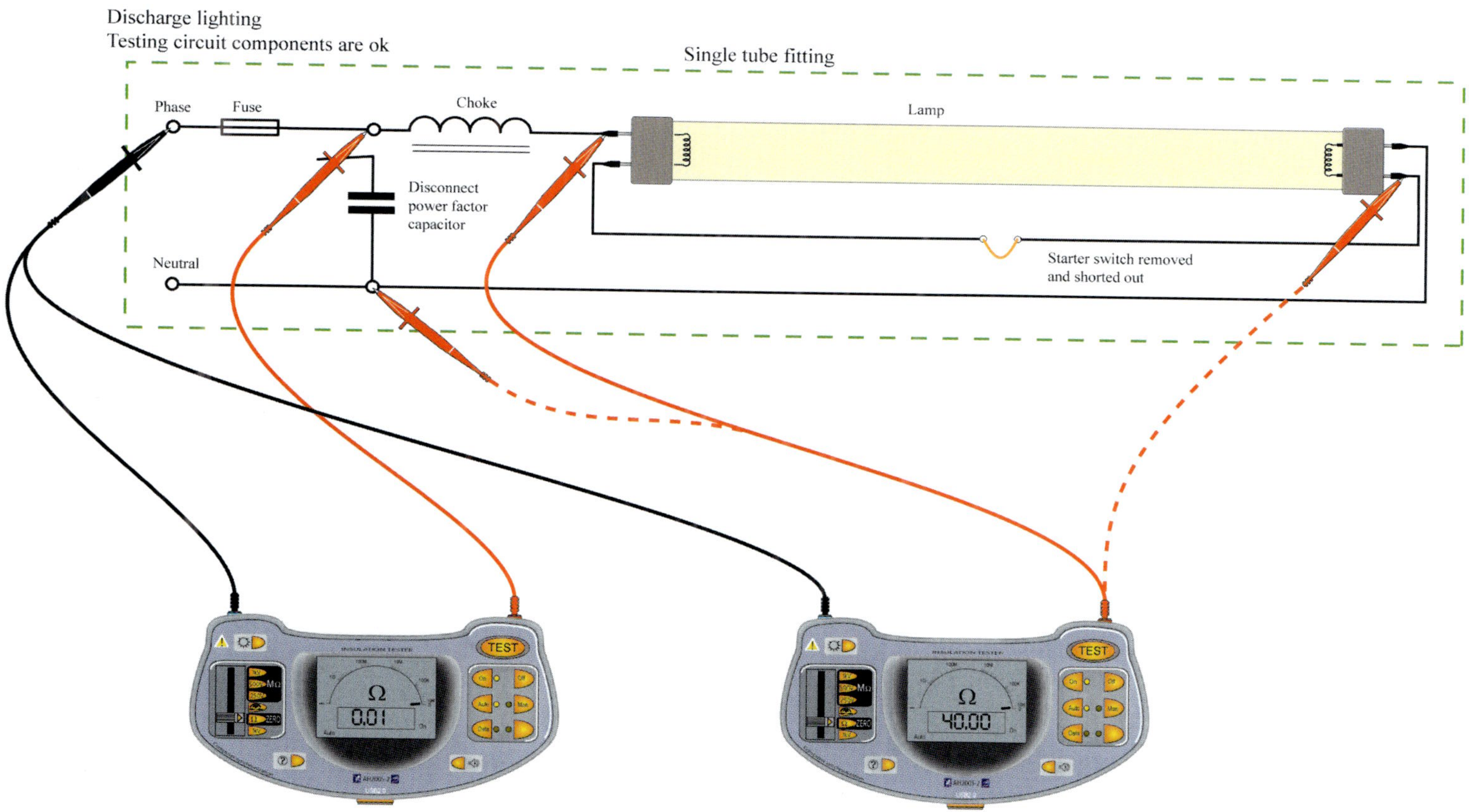

Discharge lighting
Testing circuit components are ok
Single tube fitting
Phase
Fuse
Choke
Lamp
Disconnect
power factor
capacitor
Neutral
Starter switch removed
and shorted out
TEST
Ω
0.01
40.00
ZERO

Fig: 27

The values displayed on the instruments are only examples used to explain the test

TESTING COMPONENTS IN SWITCH START FLUORESCENT LIGHTING CIRCUITS FOR INSULATION RESISTANCE

The sequence is as follows:

14 Remove all loads

15 Ensure Isolation

16 Discharge the capacitor using a test lamp or high wattage lamp, then disconnect the capacitor from the circuit

17 Disconnect fitting from the circuit

18 Remove starter switch

19 Select an insulation continuity tester and set to MΩ or 500V scale

20 Place the black lead in the earthing conductor terminal in the fitting

21 Place the red lead in the phase conductor terminal in the fitting

22 Take reading and record, the value should not be less than 0.5MΩ's

If the insulation resistance reading is negligible it is likely that the choke and/or the wiring within the fitting is short circuit to earth so you will need to test the individual components to establish where the problem is.

Important note: Discharge lighting fittings employ the use of electronic starters and self start units, so it is important that you take great care when carrying out insulation testing as the voltage induced by your instrument can damage these components. (Check with the manufacturers if you are unsure)

If you obtain a reading less than 2MΩ further investigation will be required, as the insulation either on the wiring or in the choke may be breaking down.

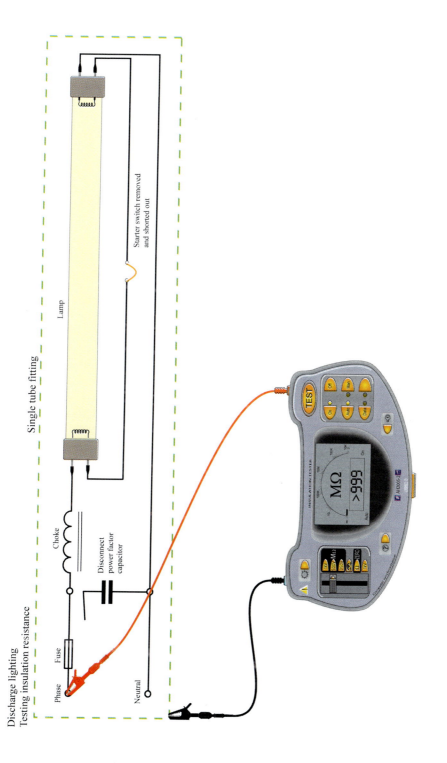

Discharge lighting
Testing insulation resistance

Single tube fitting

Phase

Fuse

Choke

Disconnect
power factor
capacitor

Neutral

Lamp

Starter switch removed
and shorted out

Fig: 28

TESTING POWER FACTOR CAPACITOR IN SWITCH START FLUORESCENT LIGHTING CIRCUITS

The sequence is as follows:

1   Remove all loads
2   Ensure isolation
3   Remove diffusers, covers and lamps
4   Pull out a little slack on the supply phase conductor and clamp current meter around conductor, (see Fig: 29A) select the appropriate range on the instrument.

The range can be determined by dividing the power by the supply voltage i.e. The following formula can be used to determine the current drawn by a twin 65w fitting.

$$\frac{2 \times 65\text{w} \times 1.8}{230\text{V}} = 1.02\text{Amps}$$

The 2 is for twin lamps, the 65w is the power of the lamps in watts, the 1.8 is a factor to allow for starting currents, and the 230V is the supply voltage.

However you will find that the calculation without the 1.8 will be closer to the current obtained, because the calculation above is a standard calculation used for calculating the current when designing the current for a discharge lighting circuit, and if I were to omit the 1.8 value it may mislead you.

5   Shield all components that will be live during the test
6   Replace, lamps
7   Switch on, take reading and record
8   Ensure isolation
9   Discharge capacitor
10   Remove capacitor lead as shown in Fig: 29B
11   Shield all components that will be live during the test
12   Switch on, take reading and record
13   Ensure isolation
14   Re-instate capacitor and all covers

Note: The value of current without the capacitor in the circuit should be higher than with the capacitor in the circuit if the two currents are the same then the capacitor is open circuit, and will need to be changed.

Testing current for confirmation
of powerfactor correction operation

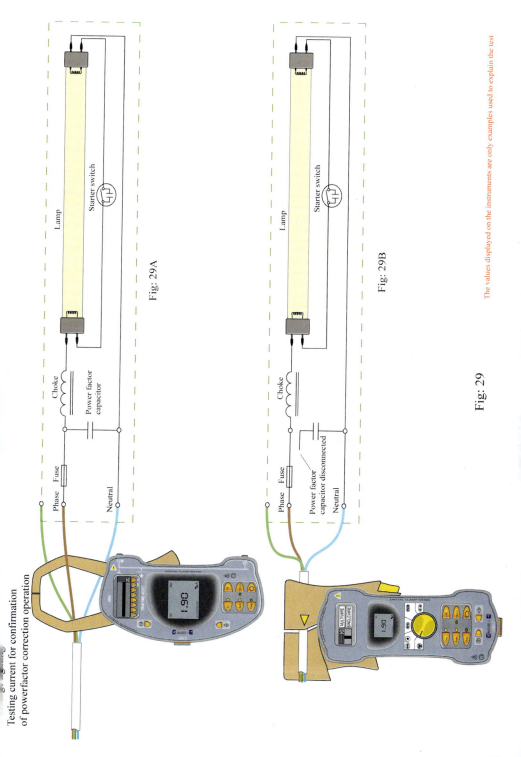

Fig: 29A

Fig: 29B

Fig: 29

The values displayed on the instruments are only examples used to explain the test

63

TESTING IMMERSION HEATERS

The sequence is as follows:

- Remove all loads
- Ensure isolation (turning off the fused spur does not constitute isolation)
- Remove cover on immersion heater
- Select an insulation continuity tester and set to Ω scale and zero
- Place the instrument on the ends of the immersion heater (See Fig: 30A)
- Record value, you should have resistance displayed on your instrument to signify that the immersion heater is continuous, if no reading is obtained check the overheat thermostate has not operated and re-set if necessary
- Now move the red probe to the other side of the thermostat as indicated, this will test the thermostat as well as the immersion, at this stage you may record $>99.9\Omega$ because the water in the cylinder may be up to temperature. If this is the case move the setting to a higher value to check the operation of the thermostat. Make note of the present setting so that it can be re-set after the test
- Now set your instrument to the $M\Omega$ or 500V scale
- Place the black lead on the earthing terminal in the immersion heater and test Phase - earth before and after the thermostat as in the previous test (See Fig: 30B)
- Take reading and record, the reading should be greater than $0.5M\Omega$

Sometimes immersions develop pinholes in the sheath of the immersion that can not be seen with the eye, causing water to penetrate the magnesium insulating material thus causing a failure in the insulation. This may not cause enough current to flow to open the circuit protective device but does constitute a failing device, and if a reading of between $0.5M\Omega$ and $2M\Omega$ is obtained, further investigation is required, as the insulation in the immersion may be breaking down.

My advice would be to change the immersion if you are in any doubt whatsoever as to its serviceability.

Please remember that it is now law that all hot water cylinders have overheat thermostates please make sure that one is fitted.

Note: It is extremely important that you not only check the serviceability of the immersion but you must check and verify that all supplementary bonding and cross bonding conductors are in place, and make absolutely sure that the immersion is wired correctly and that heat resisting insulation is applied at the terminations of the heating element.

Test: Confirmation that the thermostat and immersion heater are servicable

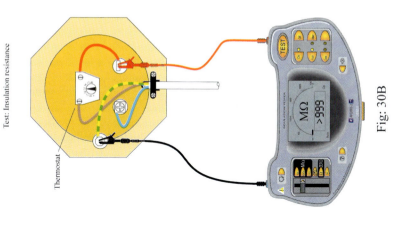

Test: Insulation resistance

Thermostat

Fig: 30B

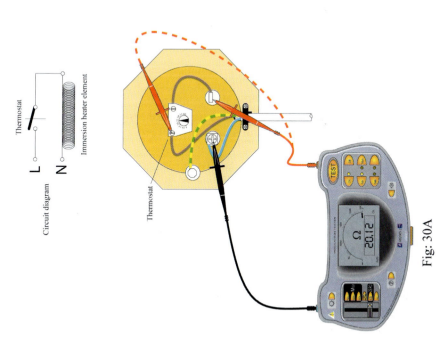

Thermostat

Circuit diagram

Thermostat

Immersion heater element

L

N

Fig: 30A

REPLACING IMMERSION HEATERS

I am aware this has nothing to do with testing, but if you have verified that the immersion heater is un-serviceable, who is going to change it?

I can assure you as long as you take care and understand what you are doing, it is fairly straightforward.

Turn off all appliances that rely on hot water to function such as washing machines.

Follow this procedure
- Locate boiler and turn off, this will stop hot water being circulated through the indirect heating coil in the cylinder
- Locate stop/gate valve, turn off water feeding the cylinder, this may be in the cupboard with the cylinder or in the loft on one of the outlets from the header tank (the larger of the two tanks in the loft)
- Open hot water taps in the bathroom to drain pipe work, the water should stop flowing fairly quickly, if it does not, the water feeding the cylinder has not been turned off. If the water can not be turned off to the cylinder tie up float valve in the header tank and drain the header tank completely
- Connect a long length of hose from the drain valve on the cylinder to the bath or WC and place old towels under the drain valve as these have a tendency to drip water
- You will need to estimate the amount of time allowed for draining down, it can take up to 30 minutes depending on where you turn off the water
- Isolate immersion heater from the electricity supply and disconnect
- Using a proper immersion spanner (rather like a big socket) gently tighten the immersion only a fraction of a turn, this will break the seal
- Unscrew the immersion gently, water will start to seep out around the screw thread if the cylinder is not drained enough (if this happens wait a little longer and then retry)
- Turn off drain valve and remove immersion, check old immersion against new one for length
- Place new immersion and fibre seal in cylinder and tighten. Do not over tighten the immersion as this can damage the cylinder (you may need to use a sealing compound such as boss white to seal the immersion)
- Reconnect immersion, refill, leaving taps turned on until water starts to flow. Finally carry out a functional test on the immersion

Note: Breaking the seal and loosening the heater should be done with water in the cylinder to prevent distortion.

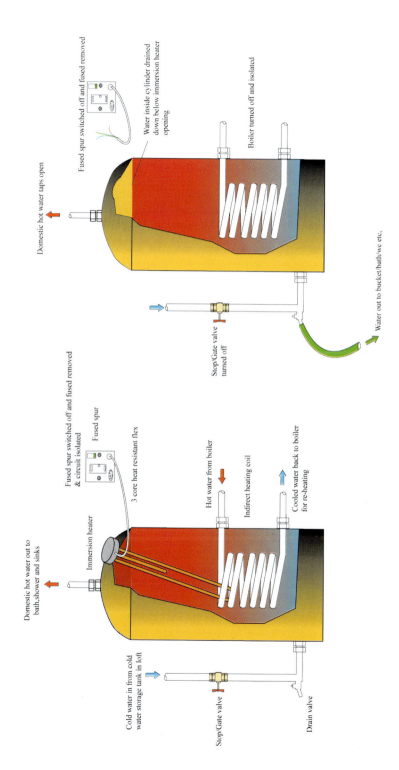

Fused spur switched off and fused removed

Water inside cylinder drained down below immersion heater opening

Boiler turned off and isolated

Domestic hot water taps open

Water out to bucket/bath/wc etc,

Stop/Gate valve turned off

Fig: 30

Fused spur switched off and fused removed & circuit isolated

Fused spur

3 core heat resistant flex

Hot water from boiler

Indirect heating coil

Cooled water back to boiler for re-heating

Immersion heater

Domestic hot water out to bath,shower and sinks

Cold water in from cold water storage tank in loft

Stop/Gate valve

Drain valve

TESTING NON ADDRESSABLE FIRE ALARMS

The subject of fire alarms is vast and I could dedicate a whole book to the subject, so I am only going to give you a brief insight into the simplest form of fire alarm, the non-addressable two wire system, which has input devices that work a little like a switch, when these are activated they "Turn On" the panel, which in turn will "Turn On" the bells, lights or sounders.

The input devices in a particular area of a building are connected to one circuit, called a zone, so that an indicator on the control panel will light to show which zone has been activated thus pointing to where the fire is located within the building.

The control panel not only displays any alarm status but it also monitors the system for faults and has controls to silence and reset the system status.

Fig:32 on page 69 is a simple fire alarm system with all input devices wired in parallel, the end of line resistor allows a small amount of current to flow in the circuit which the system control panel constantly monitors.

If this current stops flowing due to an open circuit in the cable a fault light will illuminate on the control panel signalling a fault in the circuit.

With an insulation continuity test meter set to the Ω scale we can measure the resistance in the circuit which should be equal to that of the end of line resistor, in this case $4.7k\Omega$.

If you press or activate any of the break glass switches or smoke detectors with the instrument connected as indicated at position A the resistance should change from $4.7k\Omega$ to 0.0Ω or at least a very low reading signifying that one of the input devices has reduced the resistance in the circuit sufficiently so that the control panel reacts to a fire situation and activates the alarm.

A smoke, heat detector or break glass could cause a fault similar to that which triggers a fire alarm, in order to detect this kind of fault connect the instrument as shown in position B and in a logical sequence (one after the other) progressing along the circuit disconnect each device in turn until the faulty one is located.

Remember that all types of fire alarm systems have sensitive electronic components susceptible to static electricity and extraneous voltage; take care to read the installation, maintenance and operators manuals before attempting any form of testing.

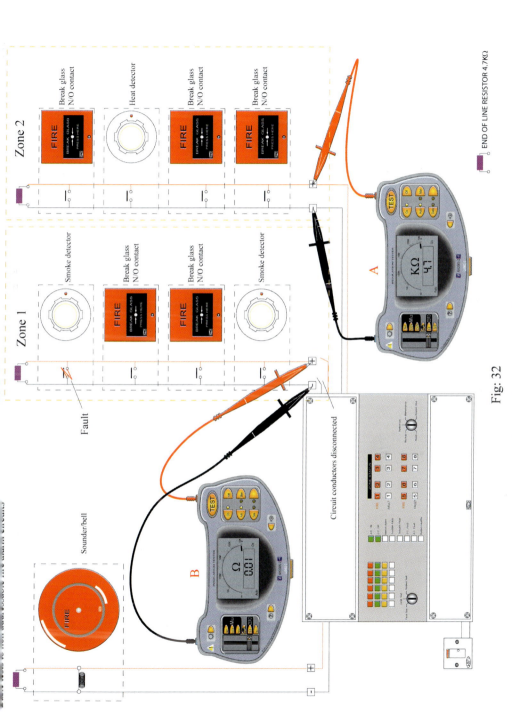

Zone 2

Break glass N/O contact

Heat detector

Break glass N/O contact

Break glass N/O contact

END OF LINE RESISTOR 4.7KΩ

Zone 1

Smoke detector

Break glass N/O contact

Break glass N/O contact

Smoke detector

Fault

KΩ 4.7

A

Sounder/bell

Ω 0.01

B

Circuit conductors disconnected

Fig: 32

69

MEASUREMENT OF EARTH LOOP IMPEDANCE (Zdb) IN A METAL CLAD 3 PHASE INSTALLATION

There has and always will be, conflicting information regarding the subject of testing earth loop impedance, especially within metalclad systems where multiple parallel paths exist.

Firstly I am going to re-educate you, and I want you to consider the earth loop impedance at a mid point distribution board as Zdb, this way we dont need to worry about parallel paths, or deciding if this is Ze or Zs, just treat it as Zdb and be done with it.

The drawing Fig:33 on page 71 depicts a typical industrial installation, where the main structure of the building consists of a steel skeleton with brick, glass and steel cladding infill and finish.

This construction method is fairly inexpensive to erect and easy to maintain and extend but does present electrical engineers with design problems and electricians with testing nightmares, especially when determining the actual Zdb at various points within the installation.

If we look at distribution board C where we have connected our earth loop tester we are measuring the earth loop impedance (Zdb) for DB3/a which consists of the following resistances.

Ze = the external impedance for the installation
(R1 + R2) resistance of phase and earth from the incoming supply to point A
(R1 + R2) resistance of phase and earth from point A to point B
(R1 + R2) resistance of phase and earth from point B to point C plus any parallel resistances such that exist at points a, b, c, d and e.

It used to be good practice to remove all parallel paths so that the actual resistance for Ze (Zdb) of each distribution board could be determined, but you can see from the very small simple installation that even here eliminating the parallel paths could be almost impossible, and even if you could would you re-instate all of the bonding links to the same degree that they were before you started the test.

It is now normal to leave all bonding conductors and links in place as the parallel paths would have been taken into account during the design, however, you must DISCONNECT THE MEANS OF EARTHING FROM THE MAIN EQUIPOTENTIAL BONDING CONDUCTORS for the duration of the test, which will mean that you have to avoid any shock hazard to the testing personnel and other people and livestock within the building during the test. See important note on page 83

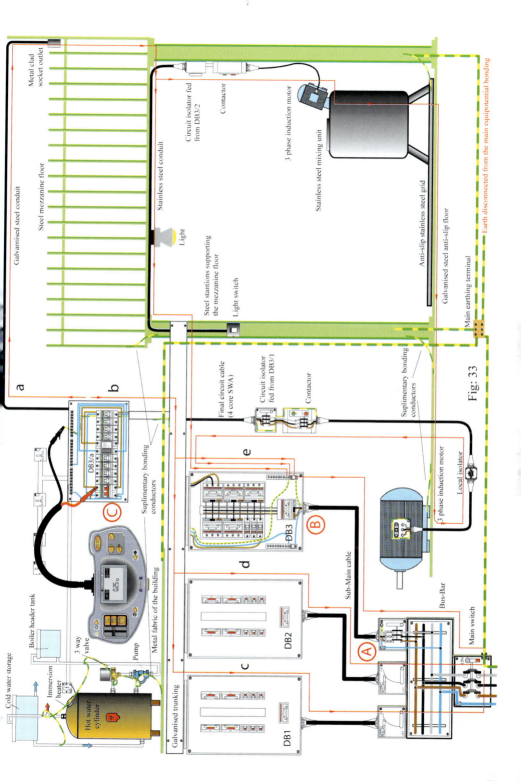

Metal clad socket outlet

Galvanised steel conduit

Steel mezzanine floor

Stainless steel conduit

Circuit isolator fed from DB3/2

Contactor

3 phase induction motor

Stainless steel mixing unit

Anti-slip stainless steel floor

Galvanised steel anti-slip floor

Main earthing terminal

Light

Steel stantions supporting the mezzanine floor

Light switch

a

b

Earth disconnected from the main equipotential bonding

Final circuit cable (4 core SWA)

Circuit isolator fed from DB3/1

Contactor

Suplimentary bonding conductors

Fig: 33

DB3/a

Suplimentary bonding conductors

e

DB3

B

3 phase induction motor

Local isolator

d

Metal fabric of the building

Boiler header tank

3 way valve

Pump

Immersion heater

Hot water cylinder

Cold water storage

Galvanised trunking

DB2

DB1

Sub-Main cable

Bus-Bar

Main switch

A

c

C

0.25Ω

MEASUREMENT OF EARTH FAULT LOOP IMPEDANCE (Zs) IN A METAL CLAD 3 PHASE INSTALLATION

With parallel paths in mind we can look at Zs for the motor circuit in the drawing Fig:34 on page 73 (by now you should be fully aware of safe isolation procedures and paper based recording systems). See important note on page 83

This is a typical industrial installation where the building consists of a steel skeleton construction with all extraneous conductive parts bonded to earth.

Our test is carried out at the remote point B by connecting the phase conductor on the meter to one of the phase conductors in the motor (L1) and the earth conductor on the meter connected to the earth terminal in the motor (R1 + R2), the supply is reinstated, the motor switched on and the test button on the meter is pressed thus displaying a reading for the actual earth loop impedance $Zs = Ze + (R1+R2)$

Lets assume a reading of 0.46Ω, now carry out safe isolation and connect the phase conductor on the meter to one of the other phases in the motor (L2) leaving the earth conductor on the meter where it was connected to the earth terminal in the motor. The supply is reinstated, the motor switched on and the test button on the meter is pressed thus displaying another reading. Note the reading and complete the test for the remaining phase (L3), all of the readings when compared should be the same as each other.

If we assume that the circuit is fed from a 10A, type B circuit-breaker, then the maximum values for earth loop impedance in table 41B2 in BS7671 is 4.80Ω, so no problem there.

However, the reading of 0.46Ω is not necessary the $Ze + (R1+R2)$ because of parallel paths. Look at the drawing and consider the red lines, which I have used to denote parallel paths for the earth to take. This means that the cpc for the motor circuit is in parallel with the mixer motor cpc at point C including the bonded metalwork in the structure of the mezzanine floor and the path through the metal clad socket on the steel framework through the galvanised steel conduit, through DB3/a, the trunking and cpc connecting all the earths together.

We can even take into account the parallel path created through the cpc for the immersion heater at point A through the bonding conductors back through the steel fabric of the building structure all of which eventually lead to the main earth bonding terminal.

Please note you should not carry out earth loop impedance testing on inverter or soft start electronic motor starters.

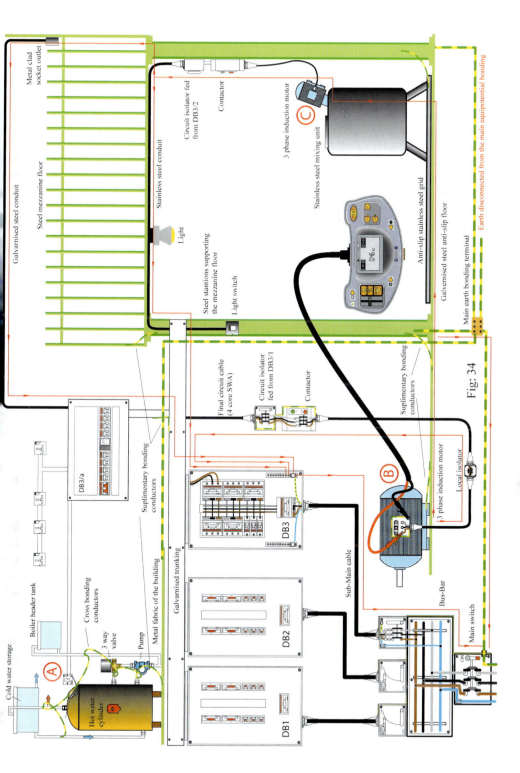

Metal clad socket outlet

Galvanised steel conduit

Steel mezzanine floor

Stainless steel conduit

Circuit isolator fed from DB3/2

Contactor

Light

3 phase induction motor

Stainless steel mixing unit

Ⓒ

Steel stantions supporting the mezzanine floor

Light switch

Anti-slip stainless steel floor

Galvanised steel anti-slip floor

Main earth bonding terminal

Earth disconnected from the main equipotential bonding

Final circuit cable (4 core SWA)

Circuit isolator fed from DB3/1

Contactor

Supplimentary bonding conductors

3 phase induction motor

Local isolator

Ⓑ

Fig: 34

Boiler header tank

DB3/a

Suplimentary bonding conductors

Metal fabric of the building

Galvanised trunking

DB3

DB2

DB1

Sub-Main cable

Bus-Bar

Main switch

Cold water storage

Ⓐ

Cross bonding conductors

3 way valve

Pump

Hot water cylinder

73

ADVANCED INSULATION TESTING

I have included this section to raise awareness of additional insulation tests that may be of value to you. However in order for us to fully understand insulation testing we need to consider some fundamentals, so lets start from the beginning.

All materials, including insulation are able to conduct electricity given a large enough potential or voltage, and contrary to belief, not all electrical current once manufactured, and sent to its destination through insulated copper conductors actually reaches its intended destination.

This loss of current is referred to as leakage or conduction current, and can be attributed to damage and imperfections within the insulation surrounding conductors and equipment, which in turn can be detrimental to the electrical circuits and equipment, possibly causing injury.

However, through routine periodic maintenance testing schedules, records are created to record the levels of leakage, which in turn can be used to determine the deterioration of the insulation over a period of time.

It's important that you understand that the integrity of the insulation can be affected by many things, such as excessive heat and cold, moisture, vibration, dirt, oil, corrosive vapours and substances, and for these reasons we should routinely test the insulation on conductors and in equipment.

We can determine the integrity of the insulation by measuring its resistance to current flow across it by applying a given voltage using an insulation or megohm continuity tester.

If a high level of resistance is measured it means that very little current is escaping through the insulation, and conversely, a low level of resistance indicates a significant amount of current may be leaking through and along the insulation.

It gets a little deep here and you may need to read this section a couple of times to become familiar with the terms.

By applying a known voltage to the conductor with an insulation tester we can measure the amount of current to earth and or between conductors passing through and along the insulation, this current is the result of three currents referred to as *capacitive current, absorption current, and leakage current.*

Capacitive current is the initial burst of current that occurs when voltage is first applied to a conductor, it generally starts out high and then drops quickly once the conductor is fully charged.

Absorption current starts out high and then drops similar to that of capacitive current, however, it falls at a much slower rate. As the voltage builds up, the absorption level in the insulation decreases. This gradual change reflects the storage of potential energy in and along the insulation. Incidentally, absorption current is an important part of the time resistance method of insulation testing.

Leakage current. Also known as conduction current, is the small steady current flowing both through and over the insulation. It goes without saying that any increase is usually an indication of the deterioration of the insulation.

Now that we have an understanding of insulation resistance and why it's important to measure it, we can look at how and when to test.

If you are serious about maintenance and creating meaningful test schedules you should accurately record the insulation resistance of new installations and equipment, because it ensures that the insulation resistance of the conductive elements in the equipment are adequately high enough to be brought into service, and is referred to as the proof test or commissioning test, and will give a reference test result by which all future testing results can be compared.

I mentioned earlier that excessive heat and cold affects the integrity of the insulation, it also alters the resistance and you need to remember that these variables will and should be taken into account when making comparative tests and readings, make a note and allowance for relative humidity and temperature.

In order to truly carryout thorough insulation testing of conductors and equipment, you will need to carry out one or more of the following tests and if the results from these tests are entered into a log over time, you create a history by which you can judge the serviceability of the equipment.

1 Proof test or commissioning test
2 Short time or spot reading test
3 Time resistance test
4 Step voltage test

The proof test is the most important test, as this initially creates the reference test results for you to compare and determine if the piece of equipment has been correctly installed.

The short time/spot reading test is carried out by connecting the tester between the conductor and earth, and applying a test voltage for 60 seconds after which an insulation resistance reading is taken and recorded. On its own as a one off test it has no value but a series of test results over several months help to build a history log.

The time resistance test can give fairly conclusive results without previous test measurements or a history log, quite helpful if you have just got involved for the first time in testing old equipment. This test method is based on taking successive readings at fixed time intervals, and then plotting the readings on a graph, it's especially effective when moisture and other contaminants might be present.

As mentioned earlier, absorption current starts out high and gradually decreases over time as voltage is applied. In a piece of equipment with healthy insulation, this trend will continue for several minutes and show an increasing level of resistance (green line on the graph). However on the other hand, if the insulation is poor, the level of resistance will flatten out after an initial burst (red line on the graph).

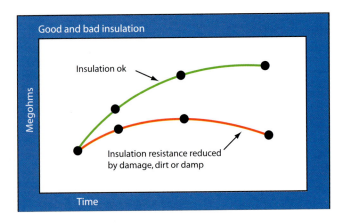

The best way to understand the results of a time resistance test is through dielectric absorption ratio, which means that we take an insulation resistance test result at 1 minute (or 10 minute) interval and divide it by a test result obtained from 30-second (or 1 minute) interval, of which the resulting value is referred to as the polarization index. (see table 1 on page 77)

In Insulation condition	1 min ÷ 30 sec interval	10 min ÷ 1 minute interval
Dangerous	-	Less than 1
Doggy	1.0 – 1.25	1.0 – 2.0
Good	1.4 – 1.6	2.0 – 4.0
Excellent	>1.6	>4.0
Please remember that environmental conditions and the state of the installation and equipment can affect your results and basic maintenance can produce more accurate results.		

Table 1

Step voltage test involves testing the insulation at two or more voltages and comparing the results. If the insulation is in good order you will see a relatively consistent resistance reading regardless of the voltage applied. On the other hand if the resistance level drops as the voltage level increases, it's usually an indication that the insulation is getting old, contaminated, or brittle and can be a result of small imperfections like pinholes and cracks that reveal themselves under increased voltages. Always start at a low voltage moving up to the higher voltages and using a test duration of not less than 1 minute.

Just remember if you are testing at the main tails of an installation the more equipment included in the test will result in a lower insulation resistance reading and for this reason it's very important to inspect the installation and understand exactly what you're including in the test.

However, if a complete installation with several pieces of equipment gives a high reading, it's safe to assume that each individual apparatus will give an even higher reading. Consequently, sometimes separating components is unnecessary and can be time consuming.

How do we know the readings are satisfactory?

Deciding what to do with the results of an insulation test can often be more complicated than actually conducting the test itself and quite often it's our own experience that tells us if a piece of equipment has a low enough insulation resistance value to warrant changing.

Every piece of equipment has an insulation resistance "personality" and no two pieces of equipment operate exactly the same, however, a safe rule is that any piece of equipment up to 1,000V should have an insulation resistance value greater than $1M\Omega$.

I know this is a bit of a cop-out but if you are in any doubt about your results you could contact the equipment manufacturers who should give you guidance and specific information on the operating characteristics, in particular to what values may be considered acceptable or questionable.

At first this may seem a little complex, but with the right equipment and some practice you will quickly get to grips with the tests, which will ultimately help you create maintenance test schedules that could truly give you an idea of the state of the equipment within you charge.

Notes

1   All references to HSE GS38 are made to raise awareness of the existence of a publication by the Health and Safety executive that provides guidance on the selection of suitable test probes, leads, lamps, voltage indicating devices and other measurement equipment used by electricians when working on or investigating power circuits with a rated voltage not exceeding 650 volts, and discusses, The dangers; The causes; Safety requirements; Systems of work.

2   **Remember to remove any and all temporary links that have been used before reinstating the supply**.

3   HSG85 titled Electricity at Work. Safe working practices, gives guidance on devising safe working practices for people who carry out work on or near electrical equipment.

4   Any isolation procedure must result in an absolute fact that the isolator supplying the piece of equipment is truly the only isolator and that no other sources of supply are likely e.g. two individual isolators supplying individual submersed pumps where flexible wiring may be crossed.

5   When removing the neutral conductor from the neutral bar, you must remember that the conductors have been terminated in a sequence, and that you should be aware that the largest load is terminated next to the main switch, which is also known as circuit 1, so the corresponding phase and cpc should also be terminated in position 1 on the phase and earthing bars in the distribution board. This will help locate all the conductors for each circuit when isolating, testing and altering the installation.

6   In order to show internal wiring I have shown covers removed in some drawings, these covers should not be removed unless you have carried out safe isolation.

7   I have referred to voltages as 400V and 230V these are to come in line with the amendments to the Regulations. (AMD 8536)

8   Discharge the capacitor by shorting the leads with a test lamp or high wattage lamp with suitable probes. (sudden discharge could cause danger)

9   Disconnecting the main earth conductor from the main earth terminal will reduce parallel paths.

10 You need to be aware that even though neon indicators in switches and equipment are not electronic devices and won't come to any harm during insulation testing, they will cause incorrect and missleading readings.

ISOLATION CERTIFICATE

ISOLATION CERTIFICATE	**Castleknight** Arcadie 173, Chemin des Hermes, 34700 SOUBES	Certificate number

This is to certify that the equipment identified below has been switched out, isolated, proved dead and warning notices posted. no attempt will be made to remove the safety devices until the Certificate has been cancelled.

SITE..

EQUIPMENT..

POINTS OF ISOLATION...

REMARKS...

Signed...Print Name...

Designation...Time...Date.............................

RECEIPT: In the case of isolation for a 3rd party		Initials
Isolation raised on behalf of.. **who hereby acknowledges the receipt of this Certificate**		Tick the box above and initial if the person raising the Isolation Certificate is also the person doing the work

Signed...Print name...

Designation...Time.................................Date................................

This section is to be used to inform the site owner/operater of an isolation, or to provide an operator or electrician with notification of isolation.

CLEARANCE & CANCELLATION

I declare that all personnel under my charge have been withdrawn and warned that it is no longer safe to work on the apparatus which was isolated as detailed above, I confirm that all equipment, locks, warning notices, tools etc, are clear and removed and that the equipment has been re-energised.

I further certify that this certificate is hereby cancelled.

Signed...................................Print name...

Designation.......................................Time...........................Date................................

Remarks...

This is an example of the type of certificate you could use.

CLEARANCE CERTIFICATE

CLEARANCE CERTIFICATE	Castleknight			
	Arcadie			
	173, Chemin des Hermes, 34700 SOUBES			

Certificate number				
Site name		Description of work		
Date	Job number			
Person in charge				
Number of workers		Is the work	Medium risk ☐	Low risk ☐
			Are Isolation Procedures Required?	**YES**	**NO**

STEPS	HAZARDS	RISK High/Med/Low	How to Reduce or Eliminate Risk	Remaining Risk

ISSUE A NEW CLEARANCE CERTIFICATE EVERY DAY DURING THE JOB IN HAND UNTIL COMPLETED UPDATE ANY ADDITIONAL HAZARDS OR CIRCUMSTANCE CHANGES

DETAILS/COMMENTS

CHECK LIST
Tick as required

Clearance Certificate fully completed ☐

Safety Method Statement available ☐

Safety Method Statement shown to site ☐

Clearance signed by contractor/poperator ☐

All obvious hazards identified and precautions taken ☐

Did a near miss occur whilst on site ☐
If yes give details

As the contractor I confirm I will carry out work as agreed

Signed.................................... Date....................................

on behalf of the site I have reviewed and agree to these work arrangements

Signed.................................... Date....................................

WORK COMPLETED SAFELY

Signature of site operator:
Comments: Date....................

This is an example of the type of certificate you could use.

DANGEROUS ELECTRICAL CONDITION NOTICE

DANGEROUS ELECTRICAL CONDITION NOTICE	Notice number

Castleknight
Arcadie
173, Chemin des Hermes, 34700 SOUBES

Contractor:	**Time:**	**Date:**

Condition..

....

....

Premises...

Address...

....

During the course of carrying out work in the above mentioned premises a condition has been discovered which constitutes a potential danger to persons, property or livestock. The condition described above should be immediately examined and any subsequent recommendations carried out or acted upon.

ADDITIONAL REMARKS...

....

....

Signed...Print name...

Designation...Time..................................Date..................................

This document is intended to be used to initially report on the existance of a potentially dangerouse electrical condition. it should not be construed to be a detailed or comprehensive list of defects that may be present within the premises and it does not take the place of an Electrical Installation Certificate or the reporting on the condition of an electrical installation such as a Periodic Inspection and Test report as required by BS7671: (2001): 2004 or any other standards.

RECEIPT

I acknowledge the receipt of this document

Signed....................................Print name...

Designation...Time.........................Date..................................

This is an example of the type of notice you could use.